Statistical Services
in Ten Years' Time

United Nations Economic Commission for Europe

Some other titles from Pergamon Press

HOUSING FOR SPECIAL GROUPS

BEHAVIOUR OF WOOD PRODUCTS IN FIRE

PROTEIN AND NON-PROTEIN NITROGEN FOR RUMINANTS

FROZEN AND QUICK-FROZEN FOOD

FACTORS OF GROWTH AND INVESTMENT POLICIES

COAL: 1985 AND BEYOND

NON-WASTE TECHNOLOGY AND PRODUCTION

THE GAS INDUSTRY AND THE ENVIRONMENT

BUILDING RESEARCH POLICIES

AIR POLLUTION PROBLEMS OF THE INORGANIC CHEMICAL INDUSTRY

HUMAN SETTLEMENTS AND ENERGY

Statistical Services in Ten Years' Time

The Operational Environment; Organization and Coordination; an 'Ideal' System; Use of Computers; Setting Priorities; Technical Developments.
Report of a Seminar of the Conference of European Statisticians, a Joint Body of the United Nations Economic Commission for Europe Washington, D.C., 21-25 March 1977

Edited by
JOSEPH W. DUNCAN
Director, Office of Federal Statistical Policy and Standards, Washington, D.C.
Seminar Chairman

Assisted by
BENNETT M. BRADY
Rapporteur

Published for the
UNITED NATIONS
by
PERGAMON PRESS
OXFORD · NEW YORK · TORONTO · SYDNEY · PARIS · FRANKFURT

U.K.	Pergamon Press Ltd., Headington Hill Hall, Oxford OX3 0BW, England
U.S.A.	Pergamon Press Inc., Maxwell House, Fairview Park, Elmsford, New York 10523, U.S.A.
CANADA	Pergamon of Canada Ltd., 75 The East Mall, Toronto, Ontario, Canada
AUSTRALIA	Pergamon Press (Aust.) Pty. Ltd., 19a Boundary Street, Rushcutters Bay, N.S.W. 2011, Australia
FRANCE	Pergamon Press SARL, 24 rue des Ecoles, 75240 Paris, Cedex 05, France
FEDERAL REPUBLIC OF GERMANY	Pergamon Press GmbH, 6242 Kronberg-Taunus, Pferdstrasse 1, Federal Republic of Germany

First edition 1978

British Library Cataloguing in Publication Data

Statistical services in ten years' time.
1. Statistical services - Congresses
I. Duncan, Joseph W II. Economic Commission
for Europe III. Brady, Bennett M
310'.7   HA12   78-40134
ISBN 0-08-022416-4

In order to make this volume available as economically and as rapidly as possible the authors' typescripts have been reproduced in their original forms. This method unfortunately has its typographical limitations but it is hoped that they in no way distract the reader.

Printed in Gt. Britain by Page Bros (Norwich) Ltd, Norwich and London.

Contents

Preface

This is a full report of a seminar hosted by the U.S. Department of State for the United Nations Economic Commission for Europe. This report contains all of the papers which were presented along with discussion summaries which were prepared from a tape recording of the conference proceedings. Bennett Brady of the Office of Federal Statistical Policy and Standards served as rapporteur for the meeting, providing the excellent discussion summaries.

Facilities at the Department of State assured a well organized meeting. Seminar arrangements were under the general direction of the Office of International Conferences. These services were deeply appreciated by all participants.

Editing of this material required some judgments concerning organization of documents and level of detail to be presented. As editor, I recognize that many useful ideas are not fully presented, but it is our hope that this report will stimulate many followup activities on the many detailed concepts which were presented.

Finally, it should be recorded that the success of this seminar should be directly attributed to the quality of discussion among the principal participants. All participants offered thoughtful contributions and the exchanges were very illuminating.

> Joseph W. Duncan, U.S.A.
> Seminar Chairman

Chief Statistician and Director
Office of Federal Statistical
Policy and Standards
U.S. Department of Commerce
United States of America

Opening Statement

B. N. Davies

Director, Statistical Division, Economic Commission for Europe

It is with great pleasure that, on behalf of the Executive Secretary of the
United Nations Economic Commission of Europe, I add my welcome to the participants in this seminar.

A great deal of work is involved in organizing an international conference
of this sort, and the bulk of it has fallen on the host country. I wish to
express my appreciation and my thanks to Mr. Duncan and his colleagues for
the efforts they have made to ensure the success of the seminar. Their
unfailing cooperation have made my own part in organizing the seminar a very
easy and pleasant one.

Many seminars have been held under the auspices of the Conference of European
Statisticians but this is the first one which from the start has been conceived as a seminar for the chiefs of the national statistical services themselves. And it has clearly attracted the chiefs. I should like to congratu-
late you, Mr. Duncan, on your success in assembling in your capital such a
galaxy of statistical talent.

The seminars serve a different purpose from the plenary sessions of the Conference of European Statisticians and the meetings of working parties held
under its programme. The purpose is to provide an opportunity for a freer
and wider-ranging discussion than is usually possible in the regular meetings.
In the seminars we are not primarily concerned with international standardization, adoption of work programmes, terms of reference of specialized meetings and similar matters and we do not need to strain to formulate recommendations in the name of the Conference. In the case of the present seminar
this is just as well since the subject under discussion hardly lends itself
to firm and precise recommendations. This is not to say that we shall be
indulging in collective gazing into a crystal ball.

There is a prospect of at least two clear results emerging from our discussions this week. First, a sharper awareness of common possibilities for
national statistical offices to exploit, and of common dangers to overcome or
avoid, in the next decade, and possibly the emergence of common views on
these matters. Second, guidance for the selection of topics to include in the
work programme of the Conference and its working parties; I expect for example that problems will be identified which the Conference will ask its
Working Party on Electronic Data Processing to study. Another point which
emerges from the papers is that while the authors have made notable efforts
to describe what is likely to happen rather than what they would like to
happen, their wishes creep in, perhaps inevitably. To some extent therefore

the discussion papers describe what the authors find adequate, or inadequate, in present statistical arrangements. In following them, and in drawing conclusions, the seminar will in effect be making a critique of present official statistics through the mirror of the future.

I believe that because of the level of the participants and the challenging nature of the subject, this seminar will be of real importance for the development of official statistics in the medium term.

Chairman's Statement

J. W. Duncan

Deputy Associate Director, Statistical Policy*, Office of Management and Budget, Executive Office of the President, United States

On behalf of the United States, I am pleased to welcome delegates and other members of delegations to the United States for this important seminar. The chief statisticians of ECE countries meet annually at the plenary sessions of the Conference of European Statisticians. Those sessions are typically filled with specific issues to be addressed and consequently it is not possible to consider the longer range structural environment for general statistical offices except in informal discussions which often occur bilaterally among members during the week of the meetings.

This seminar is designed to provide the chief statisticians of ECE member countries with an opportunity to explore common issues and to consider as a body the future directions in statistical programs.

The theme of this conference is "Statistics in Ten Year's Time." This implies a consideration of individual statistical series or concepts. However, the real focus of the meetings is the role of the Central Statistical Office in ten years' time. I am confident this will be borne out in the discussion of the various papers. For example, the opening session today focuses on the environment in which the Central Statistical Office must operate and that includes consideration of the role and mission of the statistical service in relation to the needs of policymakers and the providers of data.

The following session on the ideal statistical system and the role of computers will provide further contextual background relating to the role of the Central Statistical Office. On the third day we will consider priority setting. In fundamental terms, this topic relates to the Governmental role of the Central Statistical Office and the process by which choices are made among the many demands which are placed upon the statistical system. The discussion of technical developments on the fourth day is designed to assure that, as chief statisticians, we are all aware of the potential developments which will affect the role and functioning of the Central Statictical Office.

Finally, we have designed the last day of the conference to provide an opportunity for all chief statisticians to draw general conclusions from the previous four days' discussions. Unlike general meetings of the Conference of European Statisticians, we will not attempt to prepare an official summary

---

*In October 1977 the Statistical Policy Division became the Office of Federal Statistical Policy and Standards in the U.S. Department of Commerce. Dr. Duncan is Director of this new office.

report for each person to review at the conclusion of the conference. Rather we will have informal summaries. The full meeting is being tape recorded, and my Office will prepare a full report of the meeting which includes a summary of the discussions and the actual papers presented. A summary report of the meeting will also be prepared for presentation at the June 1977 meeting of the Conference of European Statisticians. The full report will be published by Pergamon Press in late 1978.

This is a meeting of chief statisticians of the various countries. In chairing the meetings, I will give primary recognition to the chief statisticians, and will thus restrict participation of other delegation members. This is a rare opportunity for the chief statisticians of major countries to exchange ideas fully and comprehensively. We are fortunate to have 20 chief statisticians from the ECE member countries present and to have the principal deputies representing those countries where the chief statistician was not available to participate.

Each nation operates within its unique social and constitutional setting. Traditions and current events vary widely among the countries represented at this conference. Nevertheless, it is evident by reading the materials which were prepared prior to this meeting that we do face many common issues. Our individual solutions may differ, based upon the unique circumstances of each country, but the overall challenge is so similar that the exchange of ideas should stimulate each of us, in our own ways, to enlarge our perspectives and to consider other alternatives than those which we consider in the normal course of decision making.

At the conclusion of the meetings, I expect each of us will have some new thoughts concerning the future role of the Central Statistical Office, and I further expect that we will refer to the discussions of these meetings in forthcoming sessions of the Conference of European Statisticians.

I now would like to ask Sir Claus Moser and Dr. Druchin to introduce their papers concerning the environment for statistics in the coming ten years.

The Environment in which Statistical Offices will Operate in Ten Years Time. I

Sir Claus Moser

Director, Central Statistical Office, United Kingdom

INTRODUCTION

Many of us at this Conference are able to compare life in our statistical offices today with ten years ago, and probably find it different in detail rather than in fundamental purpose or approach. I suspect the same may be true when in due course one looks back from 1987 to 1977. Of course there will be many differences in the type and range of statistics collected and in the technology used for their collection and analysis--but I doubt whether work and life in a statistical office will be <u>fundamentally</u> different from today. Whether it is or not will depend mainly on the general environment, and especially on governmental environment, in which the offices operate, and it is with this that the present paper is concerned.

To help in the preparation of these opening papers by the USSR and the United Kingdom, countries were asked to send their views and 17 did so. *Their contributions were highly informative and stimulating, but the paper, in its balance and approach, is a personal view rather than a summary of the views of others; however, all the major issues discussed were raised in one or more of the country contributions.

The comments in the paper relate primarily to developed countries and more particularly to those with market economies, whilst the parallel paper from the USSR may well concentrate more on issues prominent in centrally planned economies. In fact, there is a considerable over-lap in the issues raised in the papers from the two kinds of economies.

One other introductory remark needs to be made. In looking ten years ahead, one can either speculate on what is <u>likely</u> to happen or discuss what (in one's own view) is <u>desirable</u>. The country papers contained mixtures of the two. Here I shall <u>try as much</u> as possible to concentrate on the "likely", though some wishful thinking will creep in: it is hard to avoid the temptation of saying how official statisticians should--in one's own view--react to likely challenges. But the main emphasis is on the problems which we and our successors are likely to face in ten years time.

---

*Austria; Canada; Czechoslovakia; Finland; France; German Democratic Republic; Germany, Federal Republic of; Greece; Hungary; Italy; Poland; Romania; Spain; Sweden; Switzerland; United States of America; Yugoslavia.

THE CURRENT SITUATION

In considering the future environment, two broad facets can be identified.
One is the relation of statistical offices to the rest of government and to
other public authorities; the other is our relation with non-government groups
and individuals, both as users and suppliers of data. The former bears on
the changing role of Government itself and what, in consequence, may be
required from us; the latter bears on relations with the business community
and the public and raises problems such as privacy, confidentiality and the
burden of form-filling.

Before gazing into the future, we might try to characterize the environment
in which we work now--perhaps with a ten year backward glance in our minds.
We note the enormous growth in the scale of official statistics in most countries; it is not exceptional for the statistical office to have multiplied in
its professional complement by a factor of 5 over the last ten years, with
the grades of the top people much higher than before and with greater integration in the top councils of government. The percentage of public expenditure
spent on statistics has increased substantially, more than many other parts
of government. The range of statistics collected directly, or indirectly from
administrative data, has increased beyond recognition, accompanied by gains in
accuracy and timeliness. The technology of statistics and of ADP facilities
has improved. Decision makers are more conscious of statistics and in public
discussion statistics are more prominent. The general public, both as individuals, in various professional groupings and in business, are more sophisticated in their understanding of, and respect for, statistics, and the
standard of newspaper comment on, and use of, statistics is greatly improved.
In short, compared with ten years ago, we have more and better statistics,
they are better produced, and they are used more intelligently. If this
sounds complacent, the last year or two have removed much of the grounds for
complacency. Defects in the system have become more apparent, including prominent errors and delays, and there has been some opinion against the increasing demands made by the collectors of statistics. A more critical attitude is
evident, not only amongst policy-makers but also in the public, whose increasing participation in policy and management increases their appetite for good
data. It is against this background of increasing demand, coupled with a less
unquestioning attitude, that one looks ten years ahead.

THE GOVERNMENTAL ENVIRONMENT

The enormous increase in the demands of governments for more and better statistics has reflected the increase in government itself. Wherever one looks,
governments have tended to govern more, and it is a basic question whether
this trend is likely to continue. There are two contradictory streams of
demand. People and organizations call for more intervention and services from
government and they also express growing concern about the extent of government intervention in their lives; many take the view that governments should
govern less. It is anyone's guess to what extent, the cries for less government and less interference will carry the day against the opposite view. The
chances are that the pressure for less central government will have some success though perhaps not enough to affect the central government appetite for
official statistics; but there is no doubt it will make the climate in which
we work tougher, with more public resistance to requests for data, and thus
threats to reliability. At the same time, the activities and importance of
regional and local government is likely to increase, and with it the demand
for small area statistics.

But even if one assumes that the central government environment ten years hence will be such as to require at least as much statistical support as today, this is not to imply an unchanged official environment for statisticians. Far from it: I believe that the continued government demand for statistics will be set in a <u>very</u> changed context.

First, <u>resources</u> will not grow in parallel with demands. Public expenditure will come under increasing scrutiny and the size and growth of civil services in particular will be kept more in check, including the statistical activities. Yet the demands are certain to increase, and partly because governments themselves will need good data for setting their own priorities. An increasing conflict between demands and resources is likely, and this will mean:

(a) A greater emphasis on <u>efficiency</u> in statistical organization and production and on cost reduction <u>generally,</u> and a greater reliance on cheaper (e.g. administrative) sources.

(b) A much greater need for rational setting of statistical <u>priorities,</u> with strictly structured medium-term statistical programmes becoming the norm--related to policy needs, and with the more sophisticated use of cost-benefit techniques applied to statistics. This is a difficult area and many statistical activities are not easy to quantify in terms of benefits; but systematic costing, and the estimation of benefits wherever possible, must be the aim. In times of relatively static resources, priorities are vital--and they must encompass all parts of a statistical system.

Second, the <u>nature of government demands</u> for statistics is likely to change. This is meant in several senses:

(a) Governments attempt to steer economies--and, up to a point, social developments - with the aid of statistics. The statistical boom years have led to greater sophistication amongst policy users and to great expectations. These are often unfulfilled. The indicators are less accurate, less timely and less relevant than the policy-makers expect, and the fact that the expectations are often unrealistic may be as much the fault of the producers as of the users of statistics. Expectations and achievements must be brought closer together, with a much greater emphasis on the <u>quality</u> of the data--in terms of accuracy and timeliness.

(b) Government users of statistics will look for greater simplicity. The flood of statistics resulting from the boom years sometimes confuses or fails to help. The search will increasingly be for simple summaries, for key indicators, for simple rather than complex models.

(c) Policy-makers will increasingly want to supplement general background statistical information, with data bearing on specific problems - to help them in making particular decisions and in monitoring their consequences. Micro statistics--relating to particular areas, sectors and groups, will become more important, and over-all national statistics perhaps less so.

(d) Above all, policy-makers in government will expect from their statistical offices not so much the production of more data as, increasingly, their analysis and interpretation. The aim will be "to collect less and to use more"--and understandably so. Analytical studies, rather than mere presentations, will be expected of the statistical offices of the 1980's, with--amongst other things--advances in the statistical base for economic and social forecasts.

Statisticians will be expected to come more out into the open in assessing
their figures in terms of quality and in interpreting their meaning. Figures
will be expected to have "quality labels" attached to them. The greater
interpretative role will demand a more outward-going, politically sensitive
approach (which, one must stress, should--and can--go hand in hand with total
professional integrity).

In sum, the government environment ten years hence will be tougher as regards
resources and more critical about what official statisticians produce. The
magic of numbers may be less seductive than now. Ministers and top administrators will be sophisticated enough about statistics to want guidance about
the accuracy of the figures, as well as greater timeliness and relevance in
the figures themselves; they will want more help in analysis and interpretation. Priorities and statistical programmes will have to be explicit, with
user-orientation (for users outside and inside government) dominating their
choice. Regular routine statistics may become less important, "one-off"
surveys and analyses more so.

Trends in subject-matter are quite hard to predict. It is a safe bet that
social statistics will become even more important, as a necessary component of
increasingly systematic social planning and monitoring. The inter-relations
between different social changes will call for measurement as social policymaking becomes more sophisticated. The social consequences--often secondary
or tertiary--of economic changes will need to be analysed, and statistical
offices may be led into areas of measurement which are at present regarded as
difficult, even impossible. Statistics on distributions and differentials
will become ever more important. Generally, as societies and social policies
become less compartmentalized and more integrated, so must their statistical
analysis.

THE REGIONAL AND LOCAL ENVIRONMENT

The previous section has argued that the central government environment may
not change dramatically in the next 10 years, though the demands on the statistical services will. At 'sub-national' levels the changes are likely to be
more basic. A continued movement towards devolved and dispersed government
seems likely in many countries. The trend is towards dissatisfaction with
large units and a "small is beautiful" atmosphere will add to other pressures
for a greater spread of decision-making and administration.

This devolution of power and administrative functions from the centre to
regional, State, provincial and local authorities will substantially change
the context in which statistical offices work. It will call for improved
small-area data, and there will be a need for co-ordination to ensure that
what is collected and produced "locally" is compatible with national requirements. There will be organizational problems in how to link the statistical
operations of local offices with their central counterparts. To the extent
that the former have administrative autonomy, it will be the harder to sus-
tain an integrated statistical system and organization. This integration is
so important from the point of view of good statistics that organizational
splintering will need to be resisted, or, if it is unavoidable, accompanied
by strong co-ordinating machinery. In short, the changing 'local' environment, whilst good from the point of view of encouraging small area data, contains within it organizational difficulties. To the extent that decision and
policy making becomes increasingly de-centralized, so will the statistical

services have to be: and this could be a major change in environment.

What is certain is that sub-national authorities, whether at State, regional
or local level, will increasingly have their own statistics, research and
intelligence units, and also their own ADP installations. Networks of local
computer/data banks would be on the cards technically - with obvious advantages in supplying comparable data--but they would also raise problems of a
'political' nature.

THE PUBLIC ENVIRONMENT

The Dissemination of Government Information

Like other parts of government, national statistical offices impinge on public
life, and their relation with the public is a two-way process. On the one
hand, the public are users of government statistics, and on the other as
respondents they provide the straw with which the statistical bricks are made.
In both senses, the environment ten years hence may be very different from
now.

National statistical offices will increasingly be called upon to make the data
on which policies are based more readily available to the public at large.
Greater openness will be expected all round, partly because there will be
greater public participation in decision-making and partly because impatience
with secrecy is likely to grow. More specifically, statisticians will be
expected to make available data collected at the public's expense and through
their effort.

The public of the 1980's will be better educated and they will more consistently challenge the decisions of government. They will expect to monitor government efficiency by using official statistics and will expect readily accessible, convenient and well illustrated statistical publications, with more
guidance than now on quality of data and their meaning. Non-government users
will become more important in the thinking of statistical offices, with close
attention to public relations, the "marketing" of data, and guidance to data
sources.

For the supply of detailed data, traditional methods of dissemination will not
suffice and new technology will help. Computer terminal linkage to databanks
holding de-personalized aggregate data will become widespread and the use of
visual display units and easy-to-use analytical packages will further facilitate communication with the non-expert user. Detailed guides to the data
available in this way will be needed. Technically, such developments are
totally feasible, but they will raise serious problems of confidentiality.
In any case, the chances are that dissemination of statistics will be a mix-
ture ranging from conventional publications to data bank outputs, rather than
just the latter as might once have been expected.

It is likely that institutions other than government will increasingly analyse
and present statistics--e.g. trade unions, trade associations, stockbrokers,
research institutes. In this way, national statistical offices would lose the
monopoly they enjoy now, and there will be risks of duplication and confusion.
The more open we become with official statistics, and the better they are, the
less the risks will be. But some increase in statistical activity outside

government is inevitable: it will give the concept "official statistics" a
new meaning, and will call for strong co-ordination by the central office if
confusion is to be avoided.

Form-Filling

Whether objection to the burden of statistical enquiries placed on businesses
and individuals will harden is anyone's guess. There could be increased
resistance stemming from concern about "over-government" and objections to
authority in general. People could become more concerned about privacy, and
business firms--especially small ones--could become more impatient about
government-imposed paper work. Against this, increased openness and persua-
sion of the value of good statistics could influence attitudes in a more positive direction.

It is best to proceed on the assumption that attitudes may harden. National
statistical offices will have to ensure that society does not become, or consider itself to be, oversurveyed. Thorough scrutiny procedures will check
that all government surveys are strictly necessary and that each item of data
sought is essential. There will also be calls for national statistical
offices to control the proliferation of non-government surveys, though this
may prove difficult to implement. Educating the public about the statistical
requirements for efficient government and explaining the reasons for, and uses
of, the surveys is important. Respondents will expect to "get something
direct" out of surveys, a return for their labours. In the field of business
surveys, closer links between statisticians and company accountants will help.
The methods of data collection will change in the direction of more emphasis
on reducing the burden on the public, by e.g.:

> (a) increasing use of sample surveys (and their careful design to
> minimize sample sizes);

> (b) improving form-design;

> (c) deriving data directly from business and other organizational
> computers;

> (d) using administrative records as sources of primary data, more
> systematically and generally.

The use of compulsory surveys may decline, and therefore there will be greater
emphasis on techniques of imputation to deal with the problems of non-response. However 100 per cent surveys will still be necessary to provide data
for grossing-up procedures and for the sampling frames themselves. Regular
population censuses (asking for only a limited amount of information) will
fulfill part of this need and registers, e.g. of businesses and properties,
will be developed for use with other types of surveys.

Privacy and Confidentiality

The subject of privacy is intertwined with that of confidentiality and the
public will continue to have fears relating to both. However, subject to
fears on confidentiality being allayed, and constraints being placed on the
total burden of statistical enquiries, it is possible that public attitudes to

privacy could become more relaxed over the next ten years, and that the
limits to what one can ask, successfully, will be widened. This is hard to
gauge, and it is just as possible that citizens will increasingly challenge
the right of government to ask personal questions, and will expect stronger
justification for them. Which trend prevails will depend on attitudes to
government at the time and to complex economic and social forces. Most
likely, there will not be a consistent trend, and attitudes will vary from
country to country and from time to time.

The statistical offices will certainly have a major task in allaying the
public's fears on confidentiality at a time when there will be more pressure
for linking of data and the creation of databanks, and when all the technical
computer trends may suggest a less safe environment. Statistical laws to
cover these problems will become commonplace and, where there is no actual
legislation, there will probably be codes of conduct, with "watchdog" bodies
set up to survey their implementation. The security of personal data and the
protection of business data for individual firms will be the most important
aspects. Confidence on these and related points will be vital if statistical
offices are to retain the support of the public.

THE TECHNICAL ENVIRONMENT

The previous section touched on aspects of technology, and most countries in
their contributions discussed developments they expected in the ADP area.
Since there is a specific agenda item for this subject, only a brief mention
is needed here - enough to show that, in the view of many countries, dramatic
changes in the technical possibilities are foreseen. Mention is made of
improvements in hardware and the likely predominance of mini-computers. New
input and output technologies are foreseen, with much reference to optical
character reading, microfilms, microfiche etc. Processing is expected to
become faster and more efficient. The creation of ever larger data banks
is discussed, as are nets of databanks linking local, regional and central
offices, and even direct links with company computers. Linkage facilities
generally are expected to get much easier, facilitating the development of
integration of information systems. Administrative records are expected to
be increasingly computerized, with statistical outputs as an essential
by-product. Access to computer-held information is expected to become
easier, with facilities for "browsing" and interrogating the databanks.
"Responsiveness" of a data system held on computers is much emphasized, and
the possibilities for flexible use and access are obvious.

The potentialities are great and one can understand why some of the countries
use phrases like "computer revolution" when looking ten years ahead. But
there is, at the same time, consciousness of the danger that statisticians
will be mesmerized by ADP developments, and will be led into the construction
of increasingly complex data systems when what may be wanted is simplicity
and better use of the data already held. These dangers can be largely
avoided if statisticians have more control over ADP developments, in short
if the user voice (in this case that of statisticians) is given more weight,
and that of the technical experts perhaps less. There are also other dangers, notably in relation to privacy and confidentiality, and these too must
be given greater weight in deciding on how to develop the computer side of
our offices.

The moral seems to be to proceed with the technical developments that now
lie within our grasp, but not to be carried away by them into areas where
they might harm rather than help the service statisticians give to governments and to the public. The main need will be for statisticians to make
better, more sensitive and more relevant use of data, and to the extent that
computers help with this, they are to the good; to the extent that they distract from this aim, they should be treated with caution.

THE PROFESSIONAL AND ACADEMIC ENVIRONMENT

The professional environment for statisticians working in government may
well change. True, the main responsibility for collecting and producing
statistics will remain theirs but, if they do not rise to the challenge of
a greater analytical and interpretative contribution, this role may well be
taken over increasingly by economists and sociologists. This would reduce
statisticians to mere data-providers which would be a great pity. Professional statisticians have a great contribution to make in interpreting their
data and they should not shy away from it. Inter-action with administrators
will become better as they become more numerate, and statisticians more
literate. Inter-disciplinary teams are likely to become more commonplace.

Already the borderlines between different kinds of specialists are becoming
more blurred. Economists and statisticians already overlap in their functions, and are likely to do so increasingly. Overlaps and relationships
between statisticians and accountants, ADP specialists, OR specialists and
other groups are also likely to grow, and the breed called "statisticians"
may become less distinct and identifiable. This is not harmful as long as
the central core of the statistician's work remains clear and represents a
distinct professional contribution. At the edges, the more he becomes to
interact with other specialists and with "generalists", the better.

A very unsatisfactory aspect of the present situation is the gap between
official statisticians and the academic statistical community. This was the
main thesis in Mr. Petter Jacob Bjerve's Presidential Address to the Warsaw
session of the International Statistical Institute in 1975 and was also discussed in a paper to the American Statistical Association in 1975 by the
present author; it need not therefore be laboured here. What is apparent is
that in most countries official statisticians are relatively remote from
what goes on in university statistical departments. They make relatively
little use of advanced statistical techniques, are not in close contact with
teaching and research, and do not generally contribute to university activities; all this in a situation where official statistical work could no doubt
benefit from the help of the statistical community (as quickly becomes clear
when academic consultants are used). On the whole, university statisticians,
at any rate in mathematical and theoretical statistical departments, take
fairly little interest in official statistics, though the course work and
research in academic departments could benefit if more of it was built
around problems arising in official statistics. It is to be hoped that ten
years hence it will have been considerably narrowed. The difference between
theory and practice seems to grow each year, with increasing splintering of
the profession. Both sides--the theoreticians and the practitioners--lose
from the gap. In particular it is to be hoped that courses in statistics
will encompass a more practical outlook, with applications in official statistics having as much prominence as those in businesses, agriculture,

research and so forth. This is all the more vital if, as is argued here,
statisticians in government are to play a more analytical role: they should
emerge from their university courses with a well-balanced background. In
return, we must serve the academic fraternity better--in particular by greater
openness and access.

INTERNATIONAL ENVIRONMENT

A brief word needs to be said about the way the international environment for
our work will change. This will vary with the organizations. In some cases
their influence will be felt mainly in the search for more comparable statistical concepts and in the development of accepted international standards and
classifications. These objectives are likely to be intensified by the greater
use of automatic data processing, and there will be increased pressure for
the use of common coding systems in areas of mutual interest, e.g. documentation concerned with international trade. Just because international factors
increasingly affect national environments, the pressures from international
organizations for better and more co-ordinated data will intensify.

International organizations may increasingly sponsor the collection of new
data for the purpose of inter-country comparisons. This is particularly
likely where the organization has a formal, and partly binding, relationship
with its members (e.g. European Economic Community). Such organizations may
well become tougher in their demands, and this will cause increasing priority
choices for national offices in times of scarce resources.

The study of international problems as such--e.g. relations between economies,
markets, population movements and so forth--may grow as countries become more
intertwined, and this would intensify the demand for international and internationally comparable statistics.

CONCLUDING REMARKS

I have looked at various ways in which the environment for official statistics
may alter in the coming decade. In line with the contributions from countries, I have tried to look at the future in calm, realistic terms, avoiding
fanciful speculation. Trends likely to affect many countries, rather than a
few exceptional ones, have been emphasized.

What emerges is a challenging situation. Demands are likely to increase more
than the resources needed for satisfying them. There will thus have to be
stricter priority-setting and emphasis on efficiency. While one can assume a
continuing, perhaps a growing, respect for the role of statisticians, users
of statistics will be more critical. The statistical honeymoon of the immediate post-war decades is over. A more critical scrutiny, from within government as well as from outside, will be applied to official data, and we will be
expected to be more forthcoming about the quality of our figures and more
helpful in analysing and interpreting them. At the same time, the increasing
dispersal of decision-making will call for major improvements in small-area
data. The public generally will become more demanding, and at the same time
perhaps more resistant to data-collection and worried about privacy and confidentiality. The more open we are, the more effective in explaining why we
want data and how we plan to use them, and the better we become in disseminating statistics, the more co-operation we will get from the public.

Statistical offices will need to become more outward-looking and very user-oriented.

How statistical offices will respond to these trends will be discussed during this Seminar. There will be organizational implications (some of them discussed above) and it may be that the structure of the work force of statistical offices will change as more unskilled work is replaced by computers.

But a more fundamental organizational problem arises from the discussion in this paper. Several of the trends foreseen for the next decade imply a 'splintering' in statistical activities, away from an integrated approach: moves to more ad hoc surveys, as opposed to regular statistics, can have this effect, as can a greater stress on analysis and interpretation; the building-up of regional and local statistics can obviously have this effect; so can the development of statistical activities outside government, touched on in Section V. All these tendencies could weaken the build-up of integrated systems which has, rightly, characterized the last decade or two; and could result in a profession of autonomous cells. This would be a set-back and could undo the progress made in recent years in creating integrated and co-ordinated statistical operations. One possible outcome would be for statisticians to become confined to 'data-factories', with analysis and presentation and use done elsewhere by people with other labels. This would be highly undesirable from the points of view of integrity and of statisticians playing their full role, and must be avoided. The aim of achieving both relevance to policy and integrity, will remain one major issue to wrestle with, and the establishment of strong co-ordination to counter the effects of 'splintering' in statistical activities will be another.

The Environment in which Statistical Offices will Operate in Ten Years Time. II

A. P. Druchin

Chief, Central Statistical Board, Council of Ministers of the Russian Federation, Moscow

INTRODUCTION

General Remarks

Current and future activities of Soviet statistical bodies as well as their further development will be predetermined by economic and social progress in this country, by the needs for information on the part of Government, planning and management bodies, as well as by the interrelation among statistics and other sciences, and by the level of mechanized and automated procedures of collection and processing of statistical data.

Organizational and Structural Build-up as well as Scope of Activities of Soviet Statistical System

Perspectives and possible ways of developing the state statistical system in the Soviet Union in next ten years might be highlighted by considering the necessity to upgrade organizational and structural buildup in the scope of activity of the Central Statistical Board, USSR and its affiliations.

The Central Statistical Board, USSR and other bodies of state statistics will continue their efforts to further unify conceptual approaches and to develop uniform methods and organizational frameworks for nationwide statistics. The centralized system of recordkeeping and statistics in the Soviet Union is a mighty instrument of management, as well as for planning and coordinating the economy's development.

The Central Statistical Board, USSR will further develop ways and means to upgrade organizational frameworks, to radically improve the reporting and statistical system in the country and to secure sophisticated scientifically sound data.

The Central Statistical Board, USSR will therefore pursue the following guidelines:

Unified subject-matter development based on a scientific approach, improvement of recordkeeping, and a system of statistical indices.

Development, together with other ministries and agencies concerned, of a system of interrelated statistical indices and of a unified methodological approach; systematic upgrading of recordkeeping and statistics, further development of indices, reduction and unification of records and statistical forms; measures to secure complete consistency between statistical indices system and the planning indices system.

Collection, collation and timely presentation to the Soviet Government of scientifically sound statistical data which cover the following fields of interest:

a) fulfillment of national planning targets
b) efficiency of social production and of scientific and technological progress
c) labor productivity rise
d) proportions of industries' development
e) economic, technological and cultural progress
f) population welfare rise
g) production forces distribution
h) data which highlight available and consumed natural, labor and material resources in the country's economy
i) carrying out population censuses as well as other censuses and single time enumerations, sample surveys, etc.

Measuring the fulfillment of state planning targets. Development and implementation of statistical procedures to collect data on various industries, family budgets. Securing human environment data, natural resources statistics etc.

Study, together with the planning bodies, of current socioeconomic and technological process and phenomena through statistics; economic analysis of statistical data and presentation of reports and recommendation on the urgent problems concerning economic, social and cultural development of the country, as well as of the Union Republics and regions.

Statistical study of ways and means of increasing economic efficiency of social production and examination of the following factors affecting the economy's development: science and technology, capital investments, fixed and circulating assets, production capacities, manpower and material resources, raising technological level of production, productivity, mechanization and automation of labor-consuming processes, salaries and wages, production value, production and distribution costs, profit and rate of return, prices and price formation, setting up of economic incentives in industry, and other branches of the economy; welfare and cultural level as well as other economic and cultural problems; and disclosure of available reserves through statistical data analysis and presentation of due reports to the Government of the Soviet Union.

Supplying scientific and higher education establishments with data necessary for conducting research projects.

Publication of reports on state planning targets and other statistical materials, including statistical bulletins and statistical abstracts.

Subject-matter studies of international comparisons of statistical data including those of the Soviet Union and other countries; summary presentation of indices revealing the development levels of planned and market economy countries.

Securing broad implementation and effective use of computers and other devices to process statistical data while setting up the Automated System of Nationwide Statistics; further reduction of processing duration at minimum cost; further development and implementation of mechanized and automated processing of records in the country.

Participation in the development and setting up of the All-Union Automated System of collection and processing of data for recordkeeping, planning and management of economy development.

General guidance of primary recordkeeping in various branches of the economy.

Participation in development and perfection of unified systems of documentation, classification and coding of technological and economic information to be used in automated management systems.

Regional Statistics

Intensive efforts to develop regional statistics in the Soviet Union will continue to include structural study of the different branches of the economy as well as measurement of social phenomena throughout various territories and systems.

Regional statistics will be based on the system of basic statistical indices comprising the minimum set of indicators designed for the whole country. Otherwise regional statistics will cover some complementary indices constructed according to the economy's needs and due to the necessity of complete regional coverage of social phenomena irrespective of their nature. Establishments and institutions covered by the regional statistical system may be required to submit their data to the All-Union superior bodies or to the Union Republics administration and in some cases regional statistical data will be submitted to the local bodies only.

State Statistical Reporting System

The state statistical reporting system in the Soviet Union will, most probably, undergo no major changes, hence it will be still mainly based on reports to be submitted by all management and administration units to the bodies of state statistics on a regular basis (weekly, monthly, quarterly, annually, etc.).

Inspectorates of state statistics or district information and computing stations will have a considerable and sufficient volume of statistical information to meet the needs of local administrative bodies.

The bulk of statistical information will be still at the disposal of regional statistical departments (krajs, autonomous republics) which will get reports from industrial establishments, construction sites, collective-farms, state farms and institutions. Reports to the regional statistical departments will cover all indices necessary for statistical recordkeeping.

Statistical departments of regions (krajs, autonomous republics) will submit data to the administration bodies of their territory. Some data will be submitted to the Union Republics' Central Statistical Boards according to the data needs of the republican bodies for data.

Central Statistical Boards of the Union Republics will store, collate, compile
and submit to the Union Republics' Councils of Ministers as well as to the
Central Statistical Board, USSR adequate and scientifically sound statistics
which highlight fulfillment of plan targets as well as statistics which describe the economic and cultural development of the Union Republics.

The Central Statistical Board, USSR will submit to the Soviet Government the
entire bulk of statistics, collected and compiled throughout the country,
which show development of the national economy, distribution of production forces availability and usage of natural, material, financial and labor resources
in the country, welfare rise and cultural progress.

Planning of Statistical Projects

In planned economies the state statistical bodies may successfully perform
their functions provided a uniform plan of statistical projects is available.
All local branches and units should use a uniform schedule, methods and dates
of functioning.

Thus, the Central Statistical Board, USSR will continue developing plans of
statistical projects annually which are obligatory for all units of the state
statistical system.

This plan of statistical projects will correspond to political and management
objectives set by Government.

IMPORTANCE OF STATISTICS AND ITS ANTICIPATED INCREASE AFTER 1980

Planning of Economy's Development and Statistics

Planning of economic development is a complicated multiphase process of preparing, implementing and control of state plans fulfillment. As this process
of planning the economy's development will proceed in a uniform way, adequate
recordkeeping and statistics will be essential for its successful functioning.

All economic bodies including planning, management, financial and statistical
ones will face ever new complicated problems of scientific economic management,
of scientific planning methods and organization of economic information.

Functions of Soviet state statistics will grow more complicated and will cover
statistical recordkeeping of all development phenomena in branches of the
economy and culture. Also, they will reveal some common features of social
production development, and supply all planning and management bodies with
due reporting data. Compilation of statistical data will be performed according to scientific principles and practical guidelines.

A sharp increase in the role of statistics in the Soviet Union is anticipated
in connection with the expected development of the organizational and methodological framework to secure basic data for plan development and for necessary control and supervision over plan fulfillment.

Statistical supervision over state plans fulfillments will be further developed
to analyse the dynamic character of social production, to reveal some drawbacks with plan target fulfilment, and to find out progressive elements and to
contribute to their maximum distribution. These methods help to eliminate
those drawbacks, to uncover the causes for underfulfillment of plan targets,
to reveal the reserves to be used, to prevent the possibility of dispropor-
tions and to make due corrections to plans while they are being implemented.

Planning and statistics under planned socialist economics are closely interrelated and have common approaches. Identical indices will be used by planning
and state statistical bodies.

Further development of planning and economy's management in this country is a
major objective set by the Communist party of the Soviet Union and by the
Soviet Government for the next years. This objective immediately affects the
tasks of the Soviet state statistics. The state statistical system faces a large
and responsible task of improving record-keeping and reporting as well as
providing statistical data according to the ever increasing needs of management and planning.

To successfully solve this problem the Central Statistical Board, USSR, central statistical boards of Union republics, statistical departments of regions, autonomous republics and other statistical bodies will considerably
increase efforts to upgrade the statistical indices system so that they make
possible a comprehensive study of economic development and welfare rise in
accordance with the directives of the XXV Congress of the Communist party
of the Soviet Union. Special care will be given to development of new indices
which will help to show technological achievements in production and better
quality of goods, as well as better use of manpower, material and financial
resources, and labor productivity rise.

Soviet Statistical Science and Practice and Their Place in the System of Economic Branches of Science

Soviet economic science and its development is closely interrelated with
scientific and practical activities of state statistics. This means not only
the development of common conceptual approaches but also efforts to provide
more precise insights into processes and trends in the country's economy.
This necessity is caused by the following major factors. First, a number of
concepts which are elaborated in specific branches of economic science should
be materialized through quantitative measurements of phenomena and processes
taking place in the Soviet economic and social framework. Second, quantitative
values of processes, phenomena or actual data which help to characterize progress of this country, constitute the necessary prerequisite for due scientific and practical development.

Statistical science and practice are now very important for progress of other
economic branches and their importance will still grow after 1980. One reason
is that statistics provide actual data to be used in applied sociological,
economic and other types of research. The other reason is that no statistical progress is possible without thorough economic research or sound economic
considerations. Theoretical understanding considerably affects the framework
for statistical indices. These indices must measure various aspects of economic and social life in this country.

Quantitative values showing scope, rates and proportions of social production
development for a specific planning time span are of paramount importance for
the State Planning Committee of the Soviet Union. The Soviet state statistical system centers its attention on elaboration of scientific ways and means
to give quantitative values to phenomena and processes which are based on
certain reporting data for a specific period of time, and which are to be
used while measuring the scope, rate and proportions of Soviet economic and
social development.

SOVIET STATE STATISTICAL SYSTEM AND GOVERNMENT BODIES

Statistics in the Soviet Union will remain a powerful means of promoting the
socialist way of life, of further upgrading the level of management and of
plan oriented development of the Soviet economy. Soviet Government bodies
heavily rely on multilateral use of statistical data while 1) constructing
national development plans, 2) exercising control over plan target fulfillment
and 3) analysing social production efficiency, science and technology revolution, rise in labor productivity, distribution of productive forces, availi-
bility and use of natural, labor and material resources and increase in welfare.

The activities of Soviet state statistical bodies are highly appreciated
in this country since they make important contributions to the management
of the Soviet Union. It goes without saying that no efforts are wasted to
prove the adequacy of data collected by the Central Statistical Board, USSR
and its bodies, or "to defend" them against any allegations by other organizations. Official statistics compiled by Soviet state statistical bodies
are regarded as the only legal information source by all Government
bodies of this country; thus, those statistical data are used as a starting point while evaluating activities of establishments and institutions
as well as for planning and management of economic and cultural processes.

This status enables the Central Statistical Board of the Soviet Union to submit urgent reports to the Soviet Government and to supply planning and management bodies with major data on plan target fulfillment, on the development
of economy branches and regions. It is noteworthy that all those procedures
are performed in a very short period of time. Summary data on industrial
production in the country as well as by individual ministries for example,
are compiled by the Central Statistical Board, USSR, as of the 3rd day after
the reporting period; summary data on ploughing and harvesting by the collective and state farms are compiled on the 2nd day after the reporting period
and so on.

The next ten years will see ever more intensive efforts to reduce the time
required for processing statistical information. This reduction of data processing time will be achieved through broader usage of modern electronic data
processing equipment and of other devices and means of data transmission and
processing. An essential aspect of this trend will be further development
of a technological framework which will consist of a network of computing
centers of the Central Statistical Boards of the Union Republics and those
of regional, kraj and autonomous statistical departments as well as of district urban and rural information and computing centers.

The volume of statistical information, its subject matter and scope of application will remain closely interrelated with social development level and largely

depends on tasks set by the government of the Soviet Union.

Since all the above mentioned trends of Soviet statistical development have
been occurring during a long period of time and since they fully meet the
demands of the Government bodies, it is evident that they are liable to
remain·in the next ten years. There is evidence enough that these trends
peculiar to Soviet statistics will still be in force after 1980 considering
the fact that the need for statistical data for long-range economic and social
development as well as for decisionmaking at higher levels will grow in the
short and mid-term future.

Long-range economic and social perspectives are materialized in the Soviet
Union in five-year plans of economy development. Five-year plan objectives
are clearly defined with due corrections in annual plans of national economic development being wholly taken into account. The current five-year plan
was elaborated in accordance with the decisions of the 25th Congress of the
Communist party of the Soviet Union and covers the period 1976-1980. Long-
range national development plan is presently under way covering the period
till 1990.

Plans of national development are aimed at consequent welfare rise through
dynamic and proportional development of social productive efficiency, speeding
up scientific and technological progress. Successful fulfillment of national
plans predetermines the development of statistical science and practice in
the Soviet Union in next ten years.

MAJOR OBJECTIVES OF FURTHER DEVELOPMENT OF SOVIET STATISTICS IN THE 1980's

Further Development of Indices of Construction Methods and Improvement of Economic Analysis of Social Production Efficiency

Developing more intricate methods of indices construction and of raising the
level of economic analysis of social production efficiency is one of the most
important tasks of the Soviet state statistical bodies. Special attention
will be paid to examination of the factors affecting productivity rise, as
well as of capital assets consumption, effective use of material resources
and reduction of material, fuel and raw material consumption. Production
capacities and reduction of costs will also be closely examined.

Efficiency study, being a new field of interest of state statistics, will
be based not only on aggregate indices covering the country's economy in the
whole, but on indices for each individual branch of industry, agriculture,
transport, construction etc. This study will enable a more through insight
into social production efficiency, defining factors affecting it and revealing
some additional ways to raise efficiency. Various aspects of social production efficiency will be examined by all bodies of Soviet statistics, both
central and local.

Soviet statistics is now facing a problem of expanding and deepening economic
analysis of statistical data on available reserves to be used for successful
implementation of five-year plans. The focus of attention should be placed
on the analysis of vital economic proportions, social production structure

and those of some individual branches which are important for science and technology progress. This analysis comprises account balances as well as input-output tables.

Improvement of Balances of Construction

While elaborating national development plans, great attention should be paid to overall coverage of various balance sheets. This is caused by intensive social production increase, by the complexity of economic development plans. The Central Statistical Board of the Soviet Union is going to improve and to expand constructing value and quantity balance sheets, balances of production capacities, manpower etc. Special care will be taken to construct annual reporting balance sheets of the economy's development and input-output tables. Data available from these balance sheets will be broadly used for thorough analysis of industrial structure of social production, for more precise examination of interindustrial relations, for a detailed analysis of costs and rates of return, for an evaluation of production efficiency as well as for establishing the optimal interrelationship among various industries which enables the most effective use of material, labor and financial resources.

Development of Statistics in Some Particular Branches of Economic

Industrial statistics. Industrial statistics has a task to develop methods and implement such methods of statistical reporting and analysis which might considerably improve the presentation of data pertaining to production increase gained through better use of available assets, and through more intensive and effective consumption of equipment by the establishments available or newly built. Collection and compilation of data on factors affecting plan targets fulfillment should also be provided, including data on timely initiation of industrial projects.

A new system of statistical indices will be constructed to give insights into various aspects of production efficiency, such as labor productivity, consumption of assets, materials, etc., which will be a new step towards more perfect methods of examining industrial efficiency.

Alongside with this set of new indices, a system of summary indices will be established to enable proper evaluation of the efficiency of industrial activities on the whole.

Agricultural statistics. Significant efforts will be undertaken to assure further development of statistical indices system, for broader and more complete use of available statistical information to reveal cause and effect relations, to describe processes and phenomena in agricultural production, and to evaluate new research methods liable to be developed along with traditional ones. Special care will be given to the analysis of factors affecting crop capacity and cattle productivity, as well as gross output of agriculture.

Science and technology progress has brought into life the setting up of new organizational and management forms as well as new interbranch relations; the so called agrarian and industrial cooperation. The task set before statistical bodies for the next ten years is to develop a system of indices characterizing the activities of the interbranch establishments and of agroindustrial complexes, to thoroughly examine their impact on still growing and expanding processes of specialization and concentration.

Construction statistics. Economic analysis of construction statistics will
further promote the development of indices systems necessary to indicate the
effectiveness of inputs and of construction activities alongside with the
development of statistical indices to give qualitative estimate to construction jobs. An important prerequisite will be a system complex analysis of
efficiency of inputs which will supersede individual measurements.

New indices have to be constructed which will enable evaluation of progress
and efficiency of construction in this country. This will make statisticians
continuously account for construction affecting socialist reproduction process and systematically analyse inputs used for reconstruction, modernization
of available establishments and for new construction. Input distribution
among industries as well as regional distribution should also be explored.

As economic analysis proceeds, thorough examination of input distribution
according to the tendency of more rational productive forces displacement
will gain importance. A system of indices will be constructed which will
make it possible to make a qualitative estimate and to define the technological level of newly built establishments and of those to be reconstructed in
various industries.

Expansion of Social Statistics Studies

In the next ten years Soviet state statistical bodies will considerably expand
social statistics studies which aim at a thorough exploration of social life
under particular historic conditions. Special social programs to be used
under socialist way of life will serve as guidelines related to the global
approach of the Communist party of the Soviet Union which is aimed at consequent and multilateral promotion of the socialist way of life with a steady
rise in welfare and cultural level of the population.

Soviet statistical bodies undertake a thorough examination of various aspects
of social life in the Soviet Unions. Topics of study are social framework
and social groups structure, number and composition of population, income
distribution among various social groups, consumption of goods and services,
communal and dwelling conditions, level of public health and welfare services,
education and culture and many others.

The Central Statistical Board, USSR is going to undertake in the next ten
years systematic study of social processes taking place in this country.
This will be achieved in the first line, through a better system of social
statistics indices which will make it possible to highlight the level and
life conditions of the entire population, some individual social groups and
urban and rural sections. The studies will determine the impact of the
planned social program on the level of life of various population groups.

While studying social issues, single and sample statistical surveys as well
as opinion's polls will be broadly used along with the approved reporting
system.

It is noteworthy that, along with the efforts of Soviet State statistical bodies, intensive applied research of certain social phenomena will be under-

taken by a number of research and development institutes of the Academy of
Sciences, USSR. The main purpose of this research is to collect information
to solve social problems of the developed socialist society, namely: social
planning and forecasting, better supervision of social processes, cultural
development, communist ideas and their effective propagation, etc. Sociological research and collation of data gathered by the research and development
institutes will be performed according to theoretical concepts of the statistical science.

Economic and Statistical Studies of Human Environment

In the next ten years Soviet statistical bodies are going to considerably
expand the scope of economic and statistical research of the human environment. The Government of the Soviet Union has issued a number of bills with
the aim of settling radically some vital problems of human environment through
rational use and reproduction of natural resources. In accordance with those
decisions national development plans in the USSR will provide for measures
to build new facilities for wastes, reclamation, forest rehabilitation, flora
and fauna recovery as well as for rational use and reproduction of all natural resources.

Statistical bodies of countries which are members of the Council of Mutual
Economic Assistance have elaborated a detailed system of statistical indices
for the analysis of the human environment and rational use of natural resources. Proper statistical reports by the state statistical service supply
the governmental bodies of socialist countries with needed information on
human environment and on national plans' provisions to improve environmental
conditions. Upgrading of the human environment statistics is considered one
of most urgent problems by the Soviet state statistical bodies.

Objectives of Demographic Statistics

Socialist developments in the Soviet Union necessitate continuous examination
of problems pertaining to population, optimal relationship between the material elements of productive forces and labor forces. The next ten years will
see further development of population statistics, especially that of population censuses which are duly regarded as one of the most important sources
of data collection on number and composition of population, employment and
territorial distribution. Census returns are now used and will remain an
extremely important source for management, planning and cultural development
in the Soviet Union. They will be broadly used for various research, current
and long-range demographic estimates and forecasts.

The last All-Union census of population in the Soviet Union was conducted in
1970. The next population census is to take place in January, 1979. This
date has been determined since main population returns are to be used for
constructing five-year national development plan for 1981-1985 as well as for
long-range planning.

International recommendations will also be duly taken into account since population censuses are asked to be conducted in ten-year intervals and close
to the beginning of each ten-year period.

The schedule of the 1979 All-Union population census provides for multilateral data necessary for the analysis of demographic and social processes. Much attention will be paid, in particular, to collecting detailed information on composition and displacement of manpower, on socioeconomic employment structure, on family size and composition and on population reproduction.

The coming population census will be conducted through enquiries by enumerators who will be specially qualified for this job. The advantage of this method as compared to self-enumeration is considerable reduction of costs, speeding up of census processing and more speedy presentation of census results. A new questionnaire will serve both as technical carrier to be fed into the computer and primary document. This aggregate type of document will enable coding of some answers in the schedule by means of graphic marks which will be performed by the enumerator. Coding of complicated questions such as place of work, occupation etc. will be performed by the local statistical bodies.

Drafts of methodological and organizational guidelines to be used during 1979 All-Union population census have been tested through a pilot survey in November 1976 in nine regions with total coverage of 800,000. The results of this pilot survey will help to make necessary corrections in the schedule of the 1979 Population Census.

Development of Statistics as Brand of Science

Scientific development of statistics in the next ten years will be marked by further improvement of statistical methods of examination of mass phenomena and processes. This will be achieved through the following procedures: record-keeping upgrading, improvement of censuses and sample surveys, differential treatment of research results through grouping and subgrouping, estimates of aggregate quantitative values, evaluation of interrelations and interconnections of indices which will make it possible to reveal new typical properties and trends. Still greater attention will be paid to applied economic and statistical research and thus provide highly quantitative analysis which is closely connected with qualitative values of examined phenomena under particular historic special and time conditions.

There will be further strengthening of statistical ties with other social sciences using statistical data not only for illustrative purposes but for elaboration of theoretical concepts.

Statistics will be more closely related to other economic sciences, first of all, to planning of national development by assuring various types of economic and statistical data necessary for current and long-range plans compilation.

While developing its proper research methods, statistics will also widely apply other sciences methods. To study the quantitative aspects of social phenomena, statistics will proceed with wide application of mathematical data processing methods. Alongside with mathematical statistics such procedures will be applied as linear and dynamic programming, variation analysis, etc. These procedures will find a still broader application as a result of using modern computers.

B

IMPROVEMENT OF STATISTICS DUE TO BETTER MECHANIZATION AND AUTOMATION OF STATISTICAL DATA COLLECTION AND PROCESSING

Setting up an Automated System of Nationwide Statistics in the USSR

Further development of the Soviet state statistics will be accompanied by intensive mechanization and automation of statistical data collection and processing. An important event will be setting up the Automated System of Nationwide Statistics in the USSR which will constitute a part of the All-Union Automated System for Collection and Processing of Information for record-keeping, planning and management.

The technical buildup of the Automated System of Nationwide Statistics in-cludes available and newly set up computer centers of the Central Statistical Board, USSR, Central Statistical Boards of the Union Republics, statistical bodies of Autonomous Republics, regions, urban and rural district information and computing centers equipped with computers and other computing facilities.

The Automated System of Nationwide Statistics, connecting all the computing centers of all the statistical bodies of the Soviet Union will make it possi-ble to integrate statistical data collection and processing in this country and to ensure its multilateral application. This will bring about expansion of economic information analysis and broader application of economic and mathematical methods and models. All essential processing procedures will be automated including data transmission, storage, scanning and output. The Automated System will serve as a starting point for further centralizing sta-tistical recordkeeping by statistical bodies and for deeper and closer inter-relation with automated systems of other ministries and agencies. According to rough estimates, implementation of the Automated System of Nationwide Statistics will reduce by two or three times the duration of statistical data processing and presentation to the users.

In the next ten years Automated System of Nationwide Statistics will expand and cover some new application fields such as applied research of methods and ways of processing and analyzing economic and statistical data which will considerably facilitate their adequate use for planning and management pur-poses. Thus, the Automated System of Nationwide Statistics, while safeguard-ing higher level automatization, is aimed at further improvement of economic analysis, improving quality of data to be submitted to superior bodies, and analytic reports on socioeconomic development, fulfillment of plan targets and providing social production efficiency increase.

Electronic data processing will progress and comprise the shift from elec-tronic data processing complexes to further expansion of functional subsystems and to data base formation including interindustrial indices common for dif-ferent subsystems.

Determination of interindustrial indices within each functional subsystem will considerably integrate statistical information with the information base of the Automated System of Nationwide Statistics. It will ensure the transition to an improved data base and increase the multilateral character of economic analysis of statistical information.

Formation of the Automated data bank within the framework of the Automated
System of Nationwide Statistics will be a prerequisite for more intricate and
complicated economic analysis and for broader application of economic and
mathematical methods. The Automated data bank helps expand the application
fields such as correlation methods and factor analysis. It will ensure a
higher analytical property of statistical data and will meet the needs of
planning and management bodies in statistical data and reference tables.

The automated data bank will use uniform information and software support,
centralized catalogues, glossaries and classifications including uniform
standards and norms.

Initiation of the Automated data bank within the Automated System of Nationwide Statistics will bring about further intensification of efforts to unify
reporting documents and to reduce the number of statistical data flows by
means of accumulating, storing and frequent use of planning, instructive data
and series. Storage of current and previous data in the Automated data bank
makes it possible to increase the degree of automation of data quality control
processes which, in combination with mathematic and statistical methods of
processing, gives a favourable base for economic analysis.

Interrelation of the Automated System of Nationwide Statistics with Other Automated Systems

Alongside with Automated System of Nationwide Statistics other automated systems are being developed in the Soviet Union, among them Automated management systems in industrial branches and Automated system of planning calculations. This requires intensive efforts to coordinate activities of the above
mentioned systems, to assure those systems are consistent from the methodological, informational and technological point of view. A major effort in this
direction will be the development and use of All-Union classification of
technological and economic information to be used in all automated systems
in the Soviet Union. The main purpose of these classifications is to ensure
information consistency of automated systems within the framework of the All-
Union System for Collection and Processing of Information for recordkeeping,
planning and management which is now being initiated. This consistency is
ensured through unified names and precise interpretation of similar positions
in local or branch oriented classifications used for Automated management systems as well through codes of All-Union classifications being used for coding
input and output information of the interrelated Automated management system.
Problems might be solved either through the direct use of All-Union classification or through translators establishing the consistency of All-Union classifications with local or agency oriented ones. This procedure takes place
at the input or output of the Automated management systems.

All-Union classifications being used for intersystem information exchange will
be vitally important for statistics since statistical bodies acquire reports
from establishments and organizations subordinate to other ministries and
agencies. This is also important because statistical bodies supply other
ministries and agencies with aggregate statistical data.

Wide application of All-Union classifications makes it possible to establish
information exchange among the automated systems through carriers and channels
of communication, or through a "computer-to-computer" system which enables

increased transmission speed and adequacy of information; as well as reducing costs of data transmission.

Effective use of All-Union classifications will be of great importance for statistical bodies since it will strengthen methodological consistency of statistical data irrespective of whether they are now available or not. Conditions will thus be created to establish uniform information bases of statistical data throughout the country.

USERS OF STATISTICAL INFORMATION IN THE 1980'S

As Soviet statistical bodies have expanded their duties and competence, a field network of statistical data users has been established. It comprises, in the first line, Government planning and management bodies. Users of statistical data are also scientific and research establishments, mass media bodies, and international statistical organizations.

No radical changes in the scope of statistical data user's are anticipated in the next ten-years. Some changes, however, are possible in the nature of particular needs in statistical data. Ever greater attention is expected to be paid to content thoroughness and quality of statistical data, as well as to processing speed and adequacy of data.

Soviet statistical bodies will have to exert efforts to improve economic analysis as well as methods of collecting and processing reporting data and thus supply Government, planning and management bodies with due statistical data on plan targets fulfillment classified by industries and branches of culture.

Upgrading statistical data level places new tasks before all departments and divisions of Soviet statistical bodies. The Central Statistical Board, USSR, central statistical boards of Union republics and local statistical bodies will promote further upgrading of statistical publications which will give deeper and more comprehensive insight into national developments.

INTERNATIONAL COOPERATION IN STATISTICS

Participation in Activities of the Permanent Commision for Statistics, Council for Mutual Economic Assistance

Further development and perfection of Soviet statistics will be conducted in close cooperation with the statistical bodies of other socialist countries in accordance with major objectives of socialist economic integration of countries which are members of the Council for Mutual Economic Assistance (CMEA).

Successful solution of cardinal problems of socialist economic integration greatly depends on the analysis of economic development of countries which are members of CMEA, of their national development rates and tempos. To fulfil this task adequate and comparable data are required.

Statistical bodies of CMEA countries dispose of statistics giving a thorough insight into their economics.

Ensuring internationally comparable statistical data is an extremely complicated task. The Permanent Commission for Statistics of the CMEA will contribute
to this task by continuing efforts to ensure international comparability of
statistical data to be used for coordinating national development plans of
member countries of CMEA.

The Commission will continue its activities to unify and ensure comparability
of statistical data in various industries, as well as of classifications,
definitions and general economic indices to be used for studying economic
development problems, such as industrial classification of the CMEA countries,
The Commission is also concerned with basic methodological guidelines of economic balances construction which is necessary to give aggregate insight into
social reproduction, a list of indices pertaining to technological progress,
methodological guidelines on population statistics and manpower, the network
of basic data on economic development rates etc.

All previously exerted efforts constitute only an initial step towards com-
plex-type information statistical system which will help highlight various
aspects of socialist economic integration processes.

Cooperation of Soviet Statistical Bodies with International Statistical Organisations

In the next ten years Soviet statistical bodies intend to actively partici-
pate in activities of UN Statistical bodies, namely: UN Economic and Social
Council, Conference of European Statisticians, UN Economic Commission for
Europe, Committee on Statistics, UN Economic and Social Commission for Asia
and Pacific Conference of African Statisticians, UN Economic Commission for
Africa, Statistical bodies of UN specialized agencies: International Labor
Organisation, UNESCO, FAO and some others, as well as UN Commission on Popu-
lation.

Further relations with international statistical bodies will strongly depend
on global peace policies, progress of international cooperation in economics,
trade, science and culture. Of great significance for international cooperation are the decisions of Conference on Security and Cooperation in Europe
which provide for better quality, expansion of volume and scope of economic
information. Further efforts to secure better quality of statistical data
are better international comparability of statistical data; timely and regular
publications issue, shorter periods of publications preparation; more har-
monized statistical nomenclatures; socioeconomic and cultural comparative
studies.

This will lead to more active participation of Soviet state statistical bodies
in activities of international statistical bodies and to more intensive
cooperation in the field of statistics.

Radical improvement of international economic, scientific and technological
cooperation is expected in the next ten years which will contribute to peace-
ful coexistence of countries with different social frameworks thus increasing
internationally comparable information exchange.

While international cooperation should be expanded and deepened in all spheres
of relations among countries, it is clearly necessary to make corrections to
programs of the UN Statistical bodies especially in relation to the develop-

ment of organizational frameworks and of assuring implementation of adopted recommendations.

CONCLUSIONS

The Government of the Soviet Union places responsible tasks before Soviet state statistical bodies and highly appreciates Soviet statistics' contribution to the country's management.

The Soviet state will be developed in the next ten years in a way similar to present trends and its progress will have its input on statistical bodies which will closely adhere to Soviet Government decisions. Moreover their current and long-range plans will be compiled with the aim of ensuring timely and accurate fulfillment of those decisions. The decisions of the XXV Congress of the Communist Party of the Soviet Union about Soviet economy's development for the next five years have been thoroughly studied by the Soviet state statistical bodies and a long-range program was adopted aimed at upgrading information support necessary for exercising control over fulfillment of the decisions, for deeper economic analysis of statistical data etc.

Since the needs for statistical data are continuously increasing fundamental research is necessary to identify possibilities of meeting those needs considering the fact that as statistical information volume increases it cannot be matched by a similar increase in the number of statisticians or by due expansion of statistical facilities.

Efforts should be encouraged to reduce duration of data collection and processing which will be conducted in the next ten years with a dual purpose: shorter periods of data processing and more thorough examination of available information.

An important task of Soviet state statistics for the next ten years will continue to be coordination of collection and processing of data with the help of automated management systems and further centralization of statistical activities. This is necessitated by the need to minimize duplication of collecting and processing information and to reduce costs of statistical processing.

Successful upgrading of collection, processing and analysis of statistical information will considerably depend on progress of computers implementation.

No attempt was made in this paper to thoroughly examine all possible aspects of the problem with a more limited purpose of making some considerations about the most vital aspects.

Valuable contributions of our colleagues from Austria, Great Britain, Hungary, German Democratic Republic, Greece, Spain, Italy, Canada, Poland, Rumania, United States, Finland, France, Federal Republic of Germany, Czechoslovakia, Sweden and Yugoslavia are thankfully acknowledged, whose guidelines and recommendations, as we hope, have been duly accounted for while preparing this paper.

Comment

W. E. Duffett

Vice President, Conference Board in Canada

My remarks fall into two parts:

(1) Points which I believe have been overlooked in the discussion so far, and which I believe to be relevant.

(2) Qualitative observations, supplementing observations made by Messrs. Moser, Druchin and other speakers.

Points Not Covered

A very important issue, as I see it, is that of inflation, with respect to:

    (a) Its continuity
    (b) The necessity of recognizing it as a matter requiring special resources of a statistical nature
    (c) Its effect on the data collection arrangements of statistical offices
    (d) Its effect on other fields such as unemployment, trade, etc.

Inflation has become institutionalized in many countries with the impact of a number of factors including (1) consistent wage pressures by trade unions, (2) apathy by business firms, and (3) rejection or weakening of wage and price controls.

Increasingly, therefore, governments will be required to make provision for inflation in their statistical systems.

The initial manifestation of this awareness will likely be the elaboration of escalation systems. The consumer price index is already widely used, but a considerable variety of price indexes may need to be developed to be used for deflation in many situations. Some of these deflators will apply to components of the existing statistical structure, others relate to outside commercial contracts and transactions. Still other needs may exist in connection with debt and other financial assets in countries in which rapid inflation continues.

One requirement will be the need for assurance that value series and price statistics are comparable--in other words that the classification systems are consistent. Classification problems are often dealt with slowly and after lengthy delays as a result of shortage of resources in statistical offices.

We are only now emerging gradually from the worst depression since the
thirties, largely set off by inflationary problems and their continuing after-
math. One area in which the statistics are often considered to have been
inadequate is the measurement of unemployment. As a result of the sophisticated sample surveys developed many years ago in the U.S.A., many countries
are in good position to understand the extent and significance of unemployment. On the other hand, there is increased interest in marginal phenomena
such as "discouraged workers" who may have more or less involuntarily left
the labor force. This is a field in which more study may be desirable, but
which has not been identified by this seminar.

Another aspect of the inflation is the massive redistribution of wealth and
income which has resulted, and efforts by government to control it. The lag
in public recognition of the seriousness of this problem may well be replaced
by a desire to understand the problem and to take action to cope with it. I
have a feeling that in many countries the data are inadequate to do so.

Basic statistical records in the form of business accounts, on which statisticians depend, have been impaired in some countries by inflation, without
appropriate and adequate adjustments by statistical offices or by the accounting profession.

Profit and inventory statistics are already known to have been seriously
affected in some countries. A systematic investigation may well be appropriate to see what other data has been affected and what adjustments are
required.

One other environmental phenomenon which may be of concern to statistical
offices is separatist tendencies or demands for a degree of statistical independence within countries. The nature of the problem varies from country to
country, but could be significant.

Observations on this morning's discussion. My comments on the very useful
ideas presented are for the most part simply qualifications and I do not find
myself in outright disagreement with what has been said. The growth of
government spending, in part the product of inflation, has indeed generated
public indignation and often outrage, which will have an impact on all government programs. The public and political image of statistical programs in
a broad sense is not such that it will be immune from these pressures.

There has been an interesting discussion on the points made so well by Sir
Claus Moser about the desirability of some analysis being associated with
the statistical process. With this I am in full accord, but the proposals
cause me some trouble. The collection and tabulation of data is becoming
everywhere a more specialized function and I do not believe it to be necessary
or practicable that persons with subject matter awareness (if this is indeed
what Sir Claus is referring to) should always be deeply involved at this
stage. Certainly the trend of thinking in the U.S. seems to be in the direction of arranging for a small number of large and well-equipped statistical
units to do this kind of work. The planning, directing and analysis of
statistical investigations would, of course, remain with the subject departments and their professional staffs. I am not clear what Sir Claus means when
he refers to professional "statisticians." Does he mean survey design specialists with advanced mathematical skills, or does he mean (as I would pre-
fer) persons included in the classification of "statisticians," who may be,
in fact, economists, sociologists or other subject matter experts? If the

latter, he can count on widespread support for his suggestions. If the former, I think his position would be regarded as an extreme one.

Government as a User of Statistics is an overwhelmingly important factor in the statistical business. Whether it will become a more constructive factor is not clear. The statistical planner needs a long-run lead from the government planner, and I remain skeptical about the desire of government agencies to engage in meaningful long-run planning. Weak governments seem to be increasingly common, with a perspective no longer than the next election. Or they may be responsive to unpredictable pressures from activist groups which move rapidly from one issue to another or from powerful institutions, such as unions, pension lobbies and the like. Inflation creates circumstances which are unfavorable to governmental long-term planning and to the deliberate setting of statistical priorities.

Sir Claus' references to resentment of form-filling could perhaps be illuminated in the United States by the work of the Paperwork Commission, mentioned by Bill Shaw, which has already performed some very constructive services by pointing out, among other things, that statisticians are responsible for only about 12 per cent of the manhour burden of all questionnaires and that these are relatively easy for respondents to deal with. A most interesting study.

Nongovernment users tend to be less noticed, at least in North America, than they deserve to be. These users are indeed recognized by Sir Claus, and their growth is likely to be considerable. My own observation is that business economists are becoming more competent and demanding. In many cases they use government statistics, supplemented by their own and by services offered by organizations such as the Conference Board, to make sophisticated macro-and micro-forecasts.

Not withstanding the deplorable detachment of some university people, the teaching of economics and economic geography from a statistical point of view is becoming much more general, following the enormous success of the work of a few innovators such as Paul Samuelson. Economics teaching is spreading rapidly in secondary schools and is even appearing, quite successfully, in primary schools, where a description and statistical approach is most appropriate.

As Sir Claus points out, there are encouraging signs of improvement in the communication of statistics in digestible forms to expert and nonexpert users. The existence of nonofficial marketing and other handbooks, consisting largely of official data, is clear evidence of the opportunity for statisticians to do much more in developing constructive contact with their users.

Comment

W. H. Shaw

Consultant, U.S.A.

It is a pleasure to take part in a seminar devoted to looking ahead rather than back. If statisticians are in any way to influence the course of statistics in the coming decade, it is none too soon to start, and it is especially pertinent to start with a review of the environment in which statistical offices will operate.

My assignment is first to review the thoughtful papers of Moser and Druchin from the vantage of a one-time participating producer, a professional consumer and a business observer of United States statistics; and second to indicate what, to me, are the more pressing external problems of the United States statistical system. If my comments seem simplistic, it should be remembered that the external world views statistics simplistically.

The greater part of my remarks will be addressed to the Moser paper. His perspective centers on issues of the market economies, while Druchin's centers on those of the planned economies. There is, as Moser notes, however, "a considerable overlap in the issues." In particular, I would note the similarity of demand and subject trends and the common problem of scarce resources.

Essentially, my views reflect differences of degree, not of kind. Both Moser and Druchin are extraordinarily perceptive regarding the shape of statistical things to come. Nevertheless, it is the responsibility of a discussant to question, not just to concur, irrespective of the excellence of what he is discussing. It is also his responsibility to be forthright.

I have specific comments on the following issues: accuracy and timeliness, collection and analysis, demand trends, burden, and privacy and confidentiality. In addition, I believe that at least for the United States, public skepticism is a major obstacle to the orderly improvement of the statistical system. Finally, as you know, our system is decentralized with coordinating responsibility lodged in the Office of Management and Budget, a part of the Executive Office of the President. I will conclude my remarks with a judgement as to the direction the United States system and the coordinating process may take in the coming ten years.

As Moser points out, many statistical indicators are less accurate and less timely than policymakers require despite much improvement in recent years. He then adds that "the fact that the expectations are often unrealistic may be as much the fault of the producers as of the users of statistics." Other than changing "may be" to "is", I agree completely with this statement. I would add, however, that statisticians have two obligations to users which

they meet only too infrequently. One, they should always insist on an answer
to the question of how accurate and how timely particular statistics have to
be to meet the policy need. Given the growing scarcity of resources, we cannot afford the luxury of statistical overkill. Second, statisticians must
learn to say "no" more often. Poor statistics are not necessarily better than
no statistics; more statistics in themselves are not necessarily better aids
to policymaking. We have an obligation to stress the limitations as well as
the value.

I strongly endorse Moser's projection that policymakers will expect statistical offices to analyze and interpret -- or more euphemistically to digest --
as well as produce. But, for the United States, I suspect he underestimates
the difficulties inherent in the contradictory aspects of the expectation.
Maintaining total professional integrity while displaying a more outward-
going, politically sensitive approach will, in the United States context, be
no mean accomplishment, to put it mildly. Moreover, even prior to meeting the
expectation of the policymakers, there are the very real difficulties of
achieving genuine interface between statisticians and the various scientific
professions--social and other. There are also the even greater difficulties
of "educating" policymakers so that they understand fully both the significance and the limitations of the basic data as well as of the analyses. And
statisticians must recognize, more than they sometimes do at present, that
even the best statistics and the best analysis neither are, nor should be the
complete input for determining complex economic and social policies.

On demand and subject matter trends, I would go further than Moser. Social
statistics will not just become more important; they will become indispensable
to the policymaking process. In fact, I would venture that economic statistics without accompanying measurements of social implications--some of which,
as Moser notes, are at present regarded as difficult, even impossible--will
become less and less relevant for policymaking.

One implication of the demand trends is a substantial increase in burden.
Here, I believe Moser is a little sanguine in assuming that more education of
the public along with greater use of sampling, administrative records, etc.,
and closer links between statisticians and company accountants will effectively counter the growing hardening of public attitudes. And controlling the
proliferation of private surveys will be extremely difficult.

In the United States, burden is a serious political issue. Businessmen, for
example, have objected strongly and more or less successfully to the cost of
filling out increasingly lengthy and complicated forms. Last year, President
Ford responded to this and other pressures by ordering a flat ten per cent
cut in statistical projects. Currently the Federal Paperwork Commission. mandated by Congress, is preparing recommendations which could materially affect
the future of our statisticl system.

Privacy and confidentiality seems also to be a more serious issue in the
United States than in many countries. It is a matter of deep concern to a
wide variety of respondents; and, further has become a rallying point for
those opposed to statistics for whatever reason. Moser has discreetly described the probable trend in this area on an "either-or" basis. In the
United States, I suspect that the trend may well be "or"; that is, that the
public will challenge increasingly the right of government to ask personal
questions and the ability--or even will--of government to maintain confidentiality.

Closely associated with the privacy-confidentiality issue in the United States is the problem of public skepticism. Paradoxically it is a mix of denigration and distrust. It ranges from the cliche of "lies, damned lies and statistics" to fairly responsible beliefs that statistics can be and are used to support any argument. The current controversy about energy statistics illustrates these beliefs.

On the distrust side, public skepticism reflects an emotional feeling that government uses statistics as much to harm as to help the individual. To a considerable extent, for example, this feeling seems to account for the relatively large undercount of blacks in the 1970 Census.

With some segments of the public, distrust stems from a belief that government manipulates data for its own ends. Whatever the manifestation, public skepticism is a major problem facing both statisticians and the users of statistics.

Finally, as I noted at the beginning of these remarks, the United States statistical system is decentralized. There are, of course, advantages to decentralization but there are also serious disadvantages, including duplication and widely varying quality standards. Minimizing such disadvantages together with giving general guidance, supporting improvements, and planning are the principal functions of the coordinating group which oversees our system.

Considering its small size and its location, somewhat down the line in the Office of Management and Budget, the group performs well, in fact extremely well. It combines exceptional talent with dedication, but it can only do so much. How much is perhaps indicated by the fact that the group is only about half the size of twenty years ago. In the same interval, expenditures on principal statistical programs alone (in real terms) have at least doubled.

The coordination situation is viewed with varying degrees of disquiet by many statisticians and social scientists. One committee, representing five professional associations, is already on record as favoring substantially greater resources and alternatives to curr⁻... organizational arrrangements. Currently, this committee is engaged in developing the specifics of its goal.

My own view is that to properly monitor and produce the statistics and the analyses flowing from the demand and subject trends projected by Moser calls not only for more and better coordination but for a reexamination of the statistical system itself. If I may venture a forecast appropriate to the purpose of this Seminar, it is that our system will be reexamined in the not-too-distant future and that the United States may opt for partial centralization.

Comment

Robert Parke

Social Science Research Council, U.S.A.

Sir Claus Moser has given us a vision of the future in which the demands upon official statisticians are expected to outstrip the resources made available to them and in which more sophisticated audiences, both governmental and public, exercise more discrimination in their consumption of official data. These audiences are expected to demand greater precision in the data, greater candor from data producers with regard to the genesis and quality of the data, more analysis, that is, more guidance as to the meaning of the data, and greater simplicity in the sense of more production of key indicators and less disgorging of unexamined counts of things. Sir Claus concludes the oral presentation of his paper by suggesting that, while recent decades have been characterized by growth, and official statisticians have been able to concentrate on the building of ever more and ever larger mechanisms for collecting increasing volumes of data, it is now time for statisticians "to improve their digestion."

Having in recent decades operated in an ingestive mode, the statistical offices are now advised to operate in a more digestive mode. I think it is worth taking a look at what such a shift may mean. The aim of the statistician operating in the ingestive mode is, it seems to me, the collection and dissemination of statistical information. The aim of the statistician in the digestive mode includes this, of course, but it also includes something more-- in this mode we seek to utilize statistics to increase knowledge and understanding of changes taking place in the economy and society. Operation in the digestive mode requires that we devote more effort than at present to the pursuit of questions which emerge from the analysis of these changes. This is likely to involve a change in our attitudes toward time. In the ingestive mode we put a high premium on the currency of information whereas in the digestive mode we are likely to place greater value on the time-depth of our information, because only with time can we see patterns of change, and because it is only when we have worked with data for a while and they have "seasoned" a bit that we have confidence that we know what they mean. Where the ingestive mode involves an emphasis on the collection and processing of data, the digestive mode involves the ordering of data, modeling, and the testing of data as measures. Where the ingestive mode puts a premium upon such skills as sampling and data processing, the digestive mode calls for greater emphasis on analytical skills and on skill in compact presentation.

There is a close tie between the needs for simplification and for analysis which Sir Claus has spoken of. The simplification that is needed is a simplification of the messages conveyed by the statistics, and that depends upon measurement and analysis. We do seasonal analyses and apply the results to employment and unemployment statistics because people who are using the statistics as indicators of the state of the economy do not want the numbers dominated by the fact that it is January and the construction business is slower or that it is June and school is out.   In addition to producing masses of population and mortality data, we use them in a model of a stationary popula- tion that gives us a measure of expectation of life.   Some of the models are complicated, and the data they use must often be gathered by complicated pro- cedures.   To take the case of employment, we learned years ago that multi- state sample designs that provide for repeat enumeration of some households provide far more satisfactory estimates for the money than do simple random samples.   We have learned that simply asking people whether they are gainfully employed is far less useful than a series of questions that begins "What were you doing last week and proceeds through elaborate skip patterns to produce a set of responses that can be coded in the computer.   Useful simplification requires an emphasis on measurement and analysis--the shaping of our statistics so that they convey precise meanings and the scrutiny of the statistics to discover those meanings.

It seems to me that much of what Sir Claus has said constitutes an endorse- ment of an indicators perspective on the role of official statistics.   By this I mean a perspective which puts a premium on the validation of statistics as measures of phenomena of interest, on the replication of measures, and on analysis, informed by economic and social theory, which seeks to order data into a structured set of series and ultimately to reduce it to a smaller set meriting regular monitoring.

This perspective imposes requirements of a nature which official statisticians unassisted by academic statisticians and by economists and other social scientists are unlikely to meet.   It is therefore a perspective that draws attention to what Sir Claus and Petter Jacob Bjerve have said about the impor- tance of heightening the involvement of academic statisticians, and extends the proposition to include the involvement of researchers from other social sciences.   From observation of the situation in the United States, I believe that it is among such people that we will find the substantive knowledge, the analytical virtuosity, and the inventiveness in measurement that are needed to supplement the skills of the statistical offices in sampling, data processing, quality control, and statistical administration.   The pages of our statistical publications already contain much evidence of the usefulness of researcher inventions; in vital and health statistics alone, I think of the life table, cohort fertility tables and measures, and measures of expected and unwanted births.   In recent years the development of loglinear analysis by academic statisticians and sociologists has enormously advanced our ability to discern the patterns in categorical data. Academic statisticians are at work on stra- tegies for analyzing data collected in panel surveys.   And so forth.

The statistical offices are, it seems to me, going to be taking more of their guidance from academic statistics and social science.   Without heightened participation from these sources, it does not appear to me that the offices ┐re likely to make substantial progress in improving their digestions.   The partnership ought to be good for social science, which becomes cumulative as it comes to focus on a few key variables based on standardized measurement and a time-series base.   And it ought to be good for official statistics for, as Philip Converse has noted, it is through analysis that our data acquire the consensual meaning that makes the presentation of the statistics a self- evident necessity and makes them laden with meaning for other related aspects of the economy and society.

Discussion

The three major influences on the environment in which statistical offices
will operate in the coming decade will be a decentralization or regionalization of statistical operations, increasing demands and decreasing resources
for statistics, and growing public resistance to completing questionnaires and
supplying information. Many of the countries, particularly those of the non-
centrally planned economies, are already experiencing problems due to these
influences.

Although the participants of the seminar might be described as pessimistic
over the future environment for statistical operations, they were optimistic
on the possibilities of statistical offices shaping and influencing this environment. But to do this statisticians cannot remain in the old system; they
must move into a new system that has been discussed, planned, and developed to
meet this environment. The chief statistical officers must take the lead in
discussing the future of statistical systems. For this reason the Seminar on
Statistics of the Coming Decade was particularly important.

Decentralization of Statistical Activities

In the next few years, regional statistics will undergo a great increase and a
decentralization of statistical activities will take place. The essential
element in dealing with this devolution of power from the center to the
regions, which is already underway in many countries, will be greater coordination from the central statistical offices. Statistical offices should not
oppose this development of new sources of statistical data, but they should
strengthen in an adequate manner their role of coordinating official statistics
and develop close links with regional offices. This will be the main task of
official statisticians in the coming years.

Denmark dealt with this decentralization of statistics by adding regional aspects, that is, local units, to their registers. They collect all the information possible, not by going to the local units, but by going to the enterprises and asking for a regional breakdown of the information. In 1980 they
plan to have registers with local units. Also legislation is pending to permit regional authorities to have the names, addresses, and activity codes to
develop their own statistics. By allowing local regional authorities to use
their figures or to request additional data, they will avoid duplication.
They do not feel threatened by a takeover by regional authorities.

Decreasing Budgets for Statistics

The participants were also optimistic in discussing ways and means of meeting
the increasing demands for statistics in a time of decreasing budgets and resources. The general feeling was that when there are new demands, the ways

and means can be found to meet these demands and to obtain the necessary resources. Consumers of statistics are so interested in keeping their statistical series and their needs for new statistics are growing so fast that they
will not permit a closing down of national statistical series or a "cutting to
the bone" of budgets for statistical series.

The solution to meeting these increasing demands during times of declining resources is indicated by a rational setting of priorities. In setting priorities statistical offices must not only answer the needs of the Government but
also the needs of the whole society. Chief statistical officers could be
assisted in this if the official statistical programs were submitted to a
greater and wider audience than is the case now. Representatives of every
social element should discuss in depth the statistical programs. In this way
the needs of users can be met and the central statistical offices freed from
criticism.

Other means of bridging this growing descrepancy between the demands and resources for statistics are: greater coordination of the content of statistical programs; use of other sources for statistics such as administrative
records; estimates based on existing statistics instead of collecting new
statistics; use of sample surveys; greater use of computers and modern technical equipment, including more modern softwear; modern methods of aggregation
and dissemination of statistics, and development of data banks.

Denmark had experienced budget cuts five or six years ago; it had been their
experience that you can live with budget cuts. They had used the idea of
having people pay for statistics. In the past four to five years their annual
receipts for paid statistics have increased between 40 and 50 percent.

Public Resistance

In discussing ways of overcoming the growing resistance and indifference of
the public to completing questionnaires, the seminar agreed that more attention must be paid to making the public familiar with the activities of the
national statistical offices and to explaining the goals and uses of the statistics. Again, the participants optimistically felt that as the public uses
statistics more and understands better the need for statistics, in the future
public opinion will tie together the use of statistics and the need for information from respondents. This will have an impact on response problems.

In addition to these three environmental influences (decentralization, declining budgets, and public resistance), the four problems which seemed to
collectively worry the countries the most were: (1) confidentiality and
privacy (a number of countries are in the midst of legislative struggles to
deal with this subject): (2) the role of adminstrative data (the views ranging from those who thought this is the most important of all future developments to others who took a lesser view): (3) analysis or "digestion" of
statistics and the role of central statistical offices with respect to analysis; and (4) the question of human resources, the way in which central
statistical offices should be staffed.

Confidentiality

Two aspects of confidentiality were discussed by the participants: how to provide more small area data while maintaining our standards of confidentiality and how to define standards for the confidentiality of enterprise data.

The increasing demands for small area data immediately brings concerns about identifiable data and confidentiality. Some of the participants did not feel that the present confidentiality standards would permit central statistical offices to provide further regional breakdowns of their data, and if statistical offices are not prepared to meet the increasing demands for small area data, then they are going to become an increasingly smaller proportion of the total information industry as it relates to user groups. Other delegates did not believe that it would be difficult to overcome the problems of confidentiality; the same techniques which allow the diffusion of data now will provide adequate safeguards so that information supplied at lower levels does not go beyond the limits of statistical secrecy.

Also, if statisticians are to maintain the same confidentiality standards for firms as for individuals, that is, they will not provide identifiable data for individual firms, then in the coming years statisticians will need to redefine statistical secrecy with respect to company data.

Administrative Records

The use of administrative records as an additional source of statistical data offers many possibilities in the future but it also creates problems in maintaining statistical quality and international comparability in the data developed from this source. The use of this data will require prior coordination with the administrators of these files, particularly to avoid the proliferation of diverse data, and the closest possible cooperation with other agencies developing and using this data.

Analysis

The participants of the seminar were divided on the argument given in Sir Claus Moser's paper for further "digestion" or analysis of statistics by national statistical offices. On the one hand, some participants agreed that statisticians should be more turned outward to the public and should meet more of their needs by providing further digestion of the raw statistical data. On the other hand, some of the participants argued that although statistical offices can be expected to process statistical data further and provide quality labels, they should be cautious in going too far in the analysis of data. They agreed that it is essential that statisticians give full description of how the data is produced, the basis for the data, the quality of the data, and how it should be digested, but they argued that statistical offices should not be led to drawing conclusions from the data which might be of a political nature. Just the fact that statisticians are interpreting the data makes it possible for people to argue that they have entered the political field; whether or not in fact they have is not important. This very suspicion can

damage a statistical institute. It is immensely important to statistical
offices to have public confidence in all political courses even if they have
to pay something for this in terms of not going far enough in the analysis of
the data.

Representatives from countries with centrally-planned economies felt that statistics could not be separated from politics. Any collection of data involves
an element of analysis, i.e. deciding what is to be collected. This is the
stage where priorities have to be established and this is the time when statisticians need help from the politicians, as well as from representatives of
the sciences and arts.

It was suggested that perhaps analysis is not so dangerous in a centralized
statistical system as in a decentralized system. In a centralized system the
minister will normally rely on his own officials for drawing political interpretations, but in a decentralized system there is a great risk that the statistician will be used by the minister as an interpreter.

Human Resources

An area of growing concern and interest is bridging the gap between the academic and state statistician, between the subject matter specialist and the
statistician and between the users and producers of statistics. The International Statistical Institute's (ISI) Committee on the Integration of Statistics, of which the U.S. Representative serves as Chairman, has as one of its
tasks pursuing solutions to bridging these gaps. The Committee will report on
its work at the next session of the ISI in New Delhi. Czechoslovakia has
formed a Federal statistical council, an advisory board whose members consist,
in addition to the president of the central statistical office, of representatives of other statistical bodies, of other central bodies, of state adminis-
ration, of the sciences, and of universities.

Another human resource problem that will increase as the use of administrative
records increases is protecting the professional skills introduced into statistics from these sources.

Ways and means of enhancing the human resources base is a task that is going
to receive more and more attention from central statistical offices. It is a
task equal to the problem of enhancing the computer base.

The Organization and Coordination of Statistical Services

C. A. Oomens

Director, Economic Statistics, Central Bureau of Statistics, Netherlands

BACKGROUND

In 1974 the Netherlands Central Bureau of Statistics, for reasons not relevant here, was forced to carry out a complete reorganization. This meant
that it was necessary for ideas on ways in which a better functioning Office
could be built up, to be discussed and applied over a shorter period than is
usual and desirable in such matters.

This paper deals with the way in which the need for integration and co-ordination of a statistical system has influenced our new organization.

The main decisions as to the reorganization have been taken and executed; the
details, specifically those referring to the co-ordinating instruments, are
in the process of development. This paper describes what has been done, with
some extrapolation; ideas which did not proceed further than the initial
planning stage are referred to as such, if at all.

The actual description concentrates heavily on the statistics in the economic
field. The availability of the SNA provided a strong impetus here.

For the reader of this paper it may be further useful to know that the present
organization of the Netherlands Bureau of Statistics comprises four directorates:

1. Methods and development;
2. Economic statistics and national accounts;
3. Social and demographic statistics;
4. Administrative management, electronic data processing.

As the organization scheme in the Appendix shows, our organization is primarily
by subject-matter; some functions are centralized, but it will be noted that
these are always placed as near as possible to the subject-matter divisions
which they serve.

A recent UN report on statistical organization[1] was available when this paper
was written.

INTEGRATION, CO-ORDINATION AND OVER-ALL PLANNING

The title of this paper suggests perhaps that the subject to be discussed is

limited to the rôle of the co-orɑinator in the organization. In fact the
subject is wider: the consequences of the desire to develop both economic
and social statistics into systems in which the elements form parts of an
integrated framework, whether they relate to integration, co-ordination or
over-all planning, are described here.

The System of National Accounts can serve as an instrument to this end while
it is hoped that the System of Social and Demographic Statistics will develop
into an equally useful tool.

The national accounts appeared in the field of statistics at a relatively late
date. In the discussions concerning its form and contents the rich experience
of a few generations of statisticians has been drawn upon. This does not
however mean that the existing economic statistics more or less automatically
fit into the newly developed system. In fact a large part of the effort of
compiling the national accounts was, and still is, spent on completing, re-
arranging and in general adjusting the economic statistics which are used in
the process.

It seems hardly necessary to say that the statisticians who produce the economic statistics cannot be blamed for this situation: that is implicit in
the statement that the national accounts appeared late. But changes are necessary: economic statistics must be redesigned in such a way that they can be
fitted into the System of National Accounts without unnecessary corrections.

Statistics, of course, serve specific purposes, apart from being building
blocks of the national accounts. In our experience so far the changes brought
about by our co-ordinating activities in the economic statistics have not
diminished their usefulness for the specific puposes, while in some cases
greater possibilities for analysis result from the fact that they now fit into
a system.

The changes just described are being introduced gradually. Even in its present early phase they have a strong, widely felt influence on the work of the
Office, an influence that we feel shall become a permanent feature as the co-
ordinating instruments are further developed.

For practical reasons the discussion on our measures and plans is introduced
here in three parts, which deal respectively with:

planning and programme co-ordination
the co-ordinating function
the consequences for the subject-matter divisions.

PLANNING AND PROGRAM CO-ORDINATION

The Bureau must present an annual budget. In order to improve its planning,
it recently introduced five-year plans covering changes in the existing programme, up-dated every two years. The basis for the yearly budget then is
formed by a selection from the five-year plan.

It was realized, of course, that with this system more attention should be
given to the way in which the programme is drawn up.

For that purpose the process of decision-making is divided into two parts:

1. allocation of resources between the four directorates
2. choice of programmes for the social and economic directorates respectively.

For phase one no improvement over existing methods has been found so far: the
allocation is carried out in a series of meeting of the directorates. As regards the second phase for the economic directorate the following procedure
was adopted:

Firstly, four sets of proposals were prepared; one, as before, by the subject-
matter divisions, the second by the national accounts divisions, the third by
methods and development directorate and the fourth by the Central Planning
Office[2].

These plans were discussed with all the contributors and finalized in meetings
usually chaired by the director of economic statistics. This approach as
changed the practice that "the process of evolution of statistics..........
reflects a strong pull to serve detailed needs in specific fields as they
arise"[3].

The emphasis on the comprehensive approach meant in fact that in setting up
our plans for the period 1977-1981, priority was given to filling the gaps
left by the method of reacting to specific needs.

THE CO-ORDINATING MACHINERY

Introduction

It cannot be taken for granted that the decision to integrate economic or
social statistics will automatically lead to the desired result. It is also
not sufficient to distribute copies of the SNA and the SSDS among the chiefs
and the staffs of the subject-matter divisions with the request to rearrange
their statistics along the lines described in these reports. The fact is
that those otherwise admirable UN publications just were not written for that
purpose.

Serious experiments with committees in which experts in national accounting
participated gave somewhat better results, but at the same time showed that
certain basic decisions for the whole system had to be taken before committees
with a necessarily limited field of action could operate with any success.
Moreover, the work-load for the national accounts experts proved prohibitive,
while much of the detail of the discussions went beyond their sphere of interest.

We consequently came to the conclusion that, even in our centralized Office,
machinery should be established to implement the plans for an integrated system.

This decision was reached slowly and only gradually became more concrete. At
first our idea was to establish one central co-ordinating office (in the
methods and development directorate). The advantage of this arrangement
seemed to be that the co-ordinators would be in a part of the Office that is

not involved in the day-to-day pressures and problems of the subject-matter divisions. Later, however, it was felt to be a disadvantage that the lines of communication between the co-ordinators and the subject-matter specialists would cut across the borderline between directorates. This increases the problems of the co-ordinators, who must convince two directors in the event that their proposals are not readily accepted by the subject-matter specialists.

Another factor created additional problems: while the task for the co-ordinator in the economic field can be rather clearly described, the definition of the corresponding task in the social and demographic field is at present less apparent.

Our aim at present is to set up two co-ordinating units; one in the economic directorate, the other in the social directorate. Overall co-ordination will be the role of a standing committee at the Directors' level. The co-ordinating division in the economic directorate was officially set up on 1 December 1976.

Tasks and Instruments

Although it has already been stated before, it seems useful to mention here again that we are in a step-by-step process of building up our system of co-ordination; with respect to parts of the programme we think we have gained enough experience to go ahead confidently, with respect to others we must still learn.

The division for co-ordination of economic statistics is at present organized in such a way that it can deal with five subjects:

    (a) business registers of enterprises and establishments
    (b) standard industrial classification and statistical units
    (c) standard nomenclature of commodities
    (d) other economic classifications
    (e) definitions.

Business Registers

The task of the subdivision which deals with business registers will eventually cover the maintenance and up-dating of registers for two statistical units:

(a) the enterprise or institution; in a few hundred cases these units are split up into smaller ones that are more homogeneous as to economic activity.

(b) the local units into which the units of group (a) can be split up.

The units described under (a) are to be used in all the statistics which describe the production-process. The register with local units will be used to serve local authorities who want to compile statistics for local uses.

Register (a) is at present operational for private enterprises and institutions with personnel; the process of extending it to government agencies and enterprises without personnel has started. The 1978 business census will form the starting point for register (b). We hope to have both registers fully operational by 1980.

The register is in our view necessary for obtaining a clear delimitation between divisions according to subject-matter. It further ascertains that identical statistical units are used where this is required and that these units are uniformly classified by sector and economic activity. The register receives its information on the statistical units to be used and their classification from various sources:

Names and addresses; outside sources (social security and chamber of commerce).

Economic activity: business censuses (see chapter 6); feed-back from subject-matter divisions; special surveys by the subdivision that maintains the registers.

Size: outside sources (social security); feed-back from the subject matter divisions.

In our experience this is sufficient for an acceptable degree of accuracy of the registers.

Apart from registration, the duties of the subdivision concerned include making sure that all the interested parties in the Bureau are given the opportunity to be consulted on the choice of units and their classification. This is no simple matter: if a small enterprise with part of its activity in manufacturing and part in wholesale trade is classified in manufacturing, the division that produces manufacturing statistics must be consulted, but so must the division that produces the statistics of wholesale trade.

The register has other important uses (see chapter 6) but the co-ordinating task was assigned to it from the very beginning.

It may be useful to say that the delineation and registration of statistical units proved to be a far more complicated task than we thought. The fact that it is done centrally, in consultation with all the divisions concerned, has led to a more systematic approach and consequently to improvement of our statistics.

The improved possibilities for analysis resulting from the uniformity of units and classifications has resulted in forms of analysis that were not possible before.

Standard Industrial Classification

The Bureau had a small unit for the classification by economic activity which worked exclusively for the Population and Business Censuses. This unit, which has been and will be further strengthened, deals with the classification according to economic activity and the related problem of the choice of statistical units. The units of the registers have been chosen after long and careful discussions with all the producers of statistics; guidelines for decisions concerning these units have also been given; the chief of the subdivision arbitrates, if necessary, where opinions on interpretation of these guidelines differ.

A new national standard classification has been designed which has been introduced in most of our economic statistics. The remainder are to follow.

Indexes are also being made; links between our national classification and
the two revelant international classifications have been worked out.

The subject-matter divisions determine the classification for the units
which they handle in their statistics, on the basis of guidelines given by
the unit described here. The central register receives the information concerned, checks with the guidelines, checks with existing information and with
the information of all the other interested parties. If there are differences
of opinion, they are discussed with the parties concerned under the chairman-
ship of the chief of the unit.

Such consultation is expected to lead to agreement. In the two years in
which this arrangement has been tried it has always been satisfactory. If
it should fail, however, the problems should be put before the chief of the
co-ordinating division and, if necessary, the director of economic statistics.

Standard Classification of Commodities

Users of our statistics, and especially those who combine information on commodities from various statistics, have often complained about the lack of
uniformity of the classifications used in the statistics concerned.

Some years ago our Bureau started looking for solutions for this problem[4].

We believe that it is possible to improve the comparability of various statistics by working towards greater uniformity of the classifications mentioned. We also believe that it will be possible, by developing criteria for
such classifications, to improve the usefulness of the statistics concerned
in economic analysis[5].

A subdivision in the co-ordinating division has been charged with developing
standard commodity classifications to be used in the statistics of production
(output, input), foreign trade and household expenditure.

Other Economic Classifications, Definitions

The work done and planned on other economic classifications covers one limited
study so far: the definitions of the institutional sectors of the SNA (SNA;
5.48-5.81) have been worked out in detail.

This information is now included in the register to be used in statistics of
income originating in the production process. Much work has been done and
more remains to be done in "translating" the SNA definitions into the language
that is used in questionnaires; as a result, the same terminology is now used
for corresponding questions in a growing number of questionnaires which the
divisions of the economic directorate send to their respondents.

Coordination in the Social and Demographic Field

The unit that will be set up for co-ordination purposes in the social direc-

torate will start its work with the implementation or if necessary design of
standard classifications referring to:

occupations
education
field of interest ("purposes" in the SNA terminology, "topics")
regions.

Insofar as these classifications are used in economic statistics, the unit
works directly with the divisions of the economic directorate.

APPROACH

Very little has been said so far about the way in which the co-ordinators
approach their task.

Firstly, the general purposes of the co-ordination programme are laid before
the specialists in the subject-matter divisions and the decision of the
directorate to pursue these purposes is announced.

Next, the co-ordinator, after having studied his subject, starts discussions
in a step-by-step procedure, giving the subject matter specialists time after
each step to consult their staffs, to let the problems sink in and to come
up with suggestions. Thus, in co-operation, the new system is built up. In
the discussion, the co-ordinators point out the inconsistencies between sets
of statistics, argue, persuade. This method amalgamates the all-important
practical experience of the subject-matter specialists with the theory of an
overall-system. If the co-ordinator in the discussons takes the initiative,
he must consult all the interested parties inside and outside the Bureau; if
others take the initiative it is then the specific task of the co-ordinator
to see to it that all interested parties are consulted.

The System of National Accounts is the main source of information and inspira-
tion, where possible supplemented by publications and papers of the IMF, the
Statistical Office of the European Communities and the Conference of European
Statisticians.

The co-ordinators when necessary report to the Director of economic statistics
if the methods described above are not sufficient for achieving their purpose.

For some of its tasks, the division deals directly with the divisions of the
Directorate of Social and Demographic Statistics.

In those cases the director concerned is specifically consulted; otherwise
procedures are the same as for the economic divisions.

CONSEQUENCES OF THE CO-ORDINATION ISSUE FOR THE SUBJECT-MATTER DIVISIONS IN THE ECONOMIC FIELD

The gradual build-up of the co-ordinating division is one important consequence of the "comprehensive approach". There are also other consequences
however which, as far as they have been recognized, will be discussed in this
chapter.

Four points will be described:

    the role and place of business censuses
    the information covered in the so-called production statistics
    the delimitation of divisions by subject-matter
    the borderline between the national accounts division and the other
    subject-matter divisions.

The Role and Place of Business Censuses

The ideal economic statistics would cover all respondents, contain all relevant information and would be published early.

In practice such a range of requirements in one set of statistics is unman-
ageable, with the result that several types of statistics have been developed,
in each of which one of the requirements mentioned dominates. For the statistics of the production process, in the Netherlands' system the first requirement mentioned (cover all respondents) is met by the so-called business
census, which is held once in ten years[6]. It covers all enterprises and
institutions, except agricultural ones, which are included in a census of
their own. The number of questions is limited to those that describe the
activity and sector (but these questions are quite detailed) and information
concerning the size of the enterprise or institution (personnel, sales). In
addition sector censuses are held when the changing situation in a certain
sector makes it desirable to reinvestigate its structure.

The second requirement (all relevant information) is met by our annual production and investment-statistics. In these statistics, all enterprises or
institutions are to be covered, small ones by sample.

Censuses and annual statistics cannot be published very early, so the third
requirement is met by monthly and quarterly statistics which however give
less complete and less detailed information than the annual statistics.

For statistics in the financial field (capital finance accounts and balance
sheet statistics in terms of the SNA), annual and monthly statistics play
the same roles as in the statistics of the production process, but our plans
so far do not include all sectors and all transactions.

The main elements of this whole system of statistics were present in our
existing statistics, so that the introduction of this threefold approach as
a systematic way of making statistics in principle presented few problems.

The actual consequences were twofold:

Our annual statistics in practically all cases are not based on samples but
limited to enterprises over a certain size (cut-off). This will gradually
be changed. In order to do this, however, the subject-matter divisions must
have access to an up-to-date register. This is needed for co-ordination
purposes also (see chapter 4 under C)

The business-censuses, as stated before, completely cover one or more sectors
of the economy. These censuses, which were so far executed by the subject-
matter divisions concerned, have been brought together in one (new) division

which has no other tasks.

In addition, the purpose of these censuses has been redefined:

> to describe the structure of the sector or sectors covered
> to provide the information required for the updating of the registers.

Centralization of the business-censuses proved to be of considerable assistance for the subject-matter divisions which were freed from a task for which they had neither the experience nor enough manpower. The newly set up central censuses division is rapidly gaining the experience needed.

The Information Covered in the So-Call "Production" Statistics

At present, our annual statistics describing the production-process (insofar as they are available) provide information on sales, purchases, changes of stocks, the wage-sum and new investments. An important part of the breakdown of other costs and profits is therefore missing. This will now be changed: the new questionnaires will cover all the detail that is required for the national accounts and for other users.

The Delimitation of Divisions by Subject-Matter

The new emphasis on an "overall"-approach has not greatly changed the list of divisions in the Bureau. The distribution of work however has been rearranged somewhat.

The statistics which describe the production process with annual, quarterly and monthly statistics have been distributed over five divisions in the economic directorate[7], in such a way that all economic activities are covered.

Each division produces these statistics for the units in the register which have their main economic activity in the field assigned to the division. Information on secondary activities is collected in a form that is chosen in consultation with the corresponding division. A system of exchange of information on secondary activities between divisions is being developed.

Thus the enterprise or institution for the statistics concerned is approached by one division; it should be noted of course that the same enterprises and institutions will be approached for other statistics (prices, external trade) by other divisions.

The Borderline Between the Tasks of the National Accounts Division and of the Other Subject-Matter Divisions

As stated above, our aim is that the subject-matter divisions should compile statistics which in principle fit without further adaptation into the national accounts. This means a shift in responsibility. We intend to approach this change with great care.

Discussions have started, for instance, between the national accounts and foreign trade divisions which must ensure that the foreign trade division will

extend its activities so that ultimately this division produces the imports
and export figures as they are required for the national accounts.

Similar discussions are planned in other fields.

We are aware of the fact that it will not be possible to avoid all differences
between collected statistics and the national accounts. Corrections will
still be necessary for depreciation allowances, for "float", sampling errors
and in general when confrontation of material from different sources shows
discrepancies.

We feel that there should be a "feed-back" to the collected statistics when
such discrepancies are encountered, but so far we have not studied this
problem further.

EARLY EXPERIENCE

It seems useful to give here a short description of the experience which was
gained in the early phases of the co-ordinating work. The demand for a
better co-ordination of economic statistics came from the statisticians themselves: from the national accounts division and also from the other subject-
matter specialists who, as the field covered by their statistics grew, repeatedly found themselves intruding into territories already occupied by their
colleagues, while conversely problems arose where nobody was willing to accept
responsibility for some particularly difficult or seemingly uninteresting
angle of the economic process.

The users of statistics outside the Bureau repeatedly complained about the
lack of uniformity of statistical definitions and classifications. In general their further reaction seems to have been to try to train specialists
who knew about the peculiarities of various statistics and so to try to work
with the statistics as they were.

Practical work on co-ordination has now been going on for a few years. Re-
actions both inside and outside the Bureau have been quite favourable, although some problems were met with. The approach to the subject-matter
specialists, as described above, worked well. Co-operation was obtained in
all cases. The problems were easiest for the divisions which are going to
develop new statistics (such as annual statistics describing the production
process in commerce, transport and other services). For other divisions basic
changes had to be made in long-established practices.

The subject-matter specialists specifically welcome the idea of working on the
basis of well-discussed generally accepted criteria even if as a result their
tasks became more difficult than before. During the process of discussion
and execution in the subject-matter divisions co-ordination "experts" emerged,
people who became a new kind of specialist, helping from inside the divisions
to study and understand the many co-ordination issues. The permanence of
the new arrangements seems to be helped by this development.

Here two, of course, there is the ever-recurring dilemma for the statistician:
having to choose between continuity or improvement. In the case described
here the choice obviously was for improvement, but the repercussions have presented more problems than usual,mainly because the scale of the operation was
extensive and the changes go deep.

Both the national accounts division and the Central Planning Office experienced great problems as a result of the changes made (the Planning Office in the work on its models). This point is mentioned here because we had to take special measures (including an increase of staff for the national accounts division) to offset at least part of the damage done.

---

[1]Statistical Organization. The Organization of National Statistical Services, a Review of Major Issues. United Nations. Economic and Social Council. E/CN.3/495.

[2]A Government agency that has no formal relation with the Statistical Office.

[3]E/CN.3/495, paragraph 13.

[4]Which of course is also discussed in meetings under the auspices of various UN agencies.

[5]Netherlands Central Bureau of Statistics. Report concerning the Standard Commodity Classification.

[6]Legislation is pending which will make it possible to hold two of these censuses in each ten-year period.

[7]For a limited number of activities (health care, education, welfare services) the production statistics are compiled in the Directorate for Social and Demographic Statistics. The activity of the coordinating division extends to those statistics.

APPENDIX

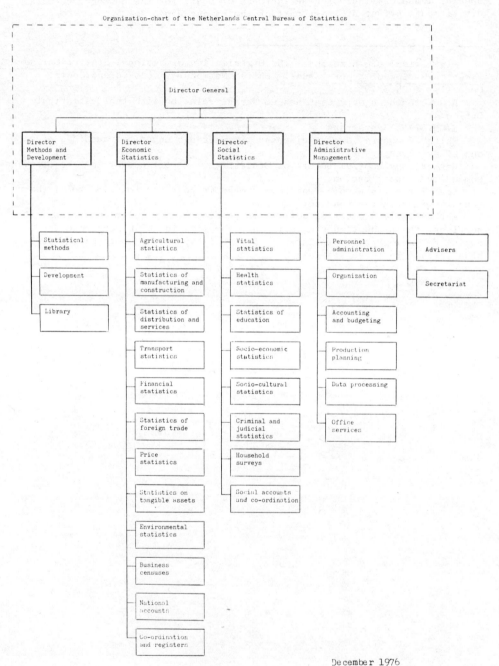

Organization-chart of the Netherlands Central Bureau of Statistics

December 1976

Discussion

The topic which drew the most attention following the presentation of the paper by the Netherlands on "Organization and Coordination" was the use of administrative records as a source of statistical data. The discussion covered the uses of these records, the extent to which statistical offices can influence their design and contents, and the problems in the use of these records.

The use of administrative records is not new; what is new is that computers make it possible to build up more and better systems. The availability of computers and other technologies will lead statisticians increasingly to use administrative records as sources of data. This is a promising development but it constitutes a real challenge in regards to the quality of the data. In order for these records to serve as a correct and satisfying source of statistical data, statistical offices are going to have to be involved in administrative operations which lead to this data.

Use of Data from Administrative Records

Administrative records have been used for many years to supplement regularly collected economic statistics and more recently in the areas of social and demographic statistics. Denmark has used the value-added tax register as the primary basis for their register of enterprises, as an indicator of economic activity in the enterprises which it covers, and as a supplement to their regularly-collected economic statistics. A number of branches of the economy, particularly in the service sector, and a lot of small enterprises are not covered in their regular collection of statistics; here, the valued added tax register is of immense value. In Sweden, administrative records are also used to supplement regularly collected statistics and to develop population, income tax, car and value added statistics. Over the past 25 years, the Bureau of the Census in the U.S. has found administrative records particularly useful in economic program areas. The principal records used, the tax records, supplement the economic census taken every five years. Administrative records for small enterprises serve as a proxy for a collected report form, thus eliminating three million small businesses from direct reporting. From these records, they develop profits, size of industry, employment, gross receipts, etc. Other by-products of these records are statistics not available from any other source and information of a sensitive nature, such as statistics on the extent of ownership of establishments by minority populations. In the social and demographic areas, administrative records are used in the evaluation of the decennial census. Statistics of births, deaths and medicare records are used to construct population models against which they can measure the coverage of the census.

In addition to providing a valuable source of data, administrative records reduce respondent burden, effectively reduce the cost of collecting data, and

C

enable countries to maintain registers of establishments between censuses.

Statistical Offices' Influence on Administrative Records

The extent to which administrative records are useful as a statistical source depends primarily on the extent to which statistical offices can exert an influence over the contents of the records. This power varies greatly among the countries. On the one end are the countries which actually house the administrative records and process the data for the program administrators; on the other end are countries which have only been able to influence the design of relatively new administrative systems.

The central statistical office of France keeps the administrative files and intervenes in the design of the administrative process to provide statistics. They also intervene in the selection of the nomenclature and the accounting principles used. This intervention insures that the files are well maintained and in accord with administrative data from other sources. The statistics will be more useful since the statisticians know what the data represents and means, and will be better adapted to the statisticians needs. In Norway, statisticians have been involved in the design of all 26 of their registers, and they thus have found them very useful for the most part. Also, they have a standardized personal identification number, thus there is a great potential for linking data although they have not done this yet.

In his paper, Mr. Omens had taken the position that he found administrative records of limited use for the purposes of the framework of national accounts. Other countries, too, noted that they had had no success in the use of admin- istrative records, or at best only limited success in the use of newly- established records in which statisticians were able to intervene in the be- ginning of development of the files. They had not been able to exert any control over existing administrative sources, nor had they had much success in influencing the output of administrative records. Attempts at influence or control were thwarted by fears, even if sometimes irrational fears, of confi- dentiality and privacy. These chief statistical officers felt that even in the future their influence would only be to a limited extent, to such things as the income definition for income taxes. They doubted whether administra- tive records would have the important meaning for the future of statistics that statisticians awaited. The concepts used in administrative records are bound to the special purposes for which they serve. Also, there are broad areas of statistics for which administrative data do not exist.

Problems With the Use of Administrative Records

Even the countries which were able to exert a large influence on the design of administrative records and thus considered them a useful source found this intervention had caused them some problems, particularly in their coordinating role. There are problems in the coordination with users of the statistics. There are conflicts of interest among users over such things as the nomencla- ture to be used, which must be dealt with. There are problems of coordination with the administrators of these files. In intervening in the design of these systems statisticians must coordinate and take into account both the statis- tical and administrative needs of the data. Also, while it may be acceptable to have a certain margin of error in statistical data, when these records are

used for administrative purposes, it is no longer acceptable or tolerable for
the individual not to be recorded or to be recorded incorrectly. These external coordination problems are going to be the most important in the years
to come, not only the coordination of administration systems, but also the co-
ordination of the blocks of total information systems in which statistics are
only one part.

There is also a problem or concern of whether the involvement with administrative records will damage the image of the central statistical office and
whether the actual housing of records which are used for administrative or
regulatory purposes in the central statistical office creates any confusion
in the office's role with regard to confidentiality. It was agreed that this
involvement will involve a risk to its public image as an agency of observation independent from the rest of government. However, France felt that it
had not been harmful to the image of the Institut National de la Statistique
et des Etudes Economiques. Luxembourg agreed and added that having one central computer for all agencies gave them a coordination position which was
helpful to them in knowing what is going on.

Other questions in regards to the use and coordination of administrative
records which were of concern to the participants were:

1. What are the organizational consequences if a central statistical office
wants to influence administrative records? Should there be a unit that deals
with administrative sources, or should the use of administrative sources best
be left to the respective subject area divisions?

2. Is it a good idea to have legislation, laws, rules, or conventions that
back-up the statistical offices' interference in administrative records?

3. Are national committees or councils helpful in getting a greater penetration into administrative sources of data?

Organization

In practice the subject matter units will take care of the use of administrative records. However, in Sweden there are so many technical coordination
problems with the computer systems that they have established one unit that
takes care of data base systems and the use of other systems.

Legislation

In general statistical offices should regard the use of administrative records as
one of their tasks and this should be mentioned in the legislation that is the
basis of their work. Statistical laws that describe the rights and responsibilities of central statistical offices should be revised to mention the use
of administrative records as their role. In Sweden, since the centralization
in the beginning of the 60's, the instructions for their central statistical
office have directed it to look after the statistical needs when administrative
records are built up. Other agencies have to consult with them when they are
developing administrative registers.

Committees

At present the use of committees in influencing the statistical uses of
administrative records have not provided much help, but they may do so in
the future.

Functional Analysis of an "Ideal" Statistical System

I. P. Fellegi*

Assistant Chief Statistician, Statistics Canada

INTRODUCTION

This paper is an attempt to present a general model for the functional analysis of statistical systems: to identify the major component processes of such a system, the inputs and outputs, the interfaces between the functions, and their control mechanisms. The purpose in so doing is to present a model which, when compared with any particular implementation, might highlight missing or weak components. Thus the system is "ideal" in the sense that all components are developed and functioning.

Several points should be emphasized.

The choice of the term "Statistical System", instead of "Statistical Office" is not accidental. No assumption is made about how the system is organized, in particular, whether it is centralized or decentralized. As far as the term "Statistical System" is concerned, different authors use it differently We use the term for purposes of this paper in the narrow sense: as the sum total of the functions carried out by statistical agencies - excluding, for example, the external users of statistics and the respondents. Perhaps "Statistical Service" might have been a more suitable term, except that we want to emphasize the functioning of this service as a system.

In the literature and general discussions that have taken place with respect to the organization of the statistical service much of the discussion relates to the issue of centralization or decentralization. Other issues influencing the particular organization are:

(i)   the political system and government organization
(ii) the social-cultural environment in the country (e.g. the public acceptance or otherwise of registers of persons)
(iii) economic structure
(iv) level of technological development
(v) the type and quality of human resources which are available.

Moreover, it is well known that any formal line organization can provide only the very crudest indication of the way the statistical service actually carries out its activities, i.e. the way it functions.

---

*With substantial inputs from G.J. Brackstone, D.J. Dodds, Mrs. M.R. Hubbard, E. Outrata and W.M. Podehl of Statistics Canada.

The foregoing suggests that because these factors are likely to be different in each country and over time, and receive different weights in organizational considerations, a fruitful general approach to understanding the functioning of the statistical service might involve a systems type of analysis of functions-- as opposed to organizational units. Once the "system" with all its components is identified, any existing organization can be analysed in the context of this system in order to identify strengths, weaknesses, gaps, etc.

The analysis proceeds in a hierarchical fashion, starting with the overall statistical system and proceeding, in successive stages, to subdivisions of it, called functions and subfunctions, but which could also be called component processes. The particular method of dividing up the functions is based on inputs and outputs. What emerges is, therefore, very much process-oriented, as opposed to, for example, organization-oriented. It highlights interrelationships, including causalities, both within the system and with the outside world.

In addition to its inputs and outputs, we tried to identify the controls for each function. It should be emphasized that the term "control" in this paper is a systemic one, as opposed to the purely management one. Thus by control of a function or subfunction is meant the mechanism ensuring that the particular process carries out its assigned function in harmony with the rest of the system. For each function (process), where appropriate, we identify checkpoints at which the control function must be exercised, as well as the tools and yardsticks whereby the control assesses each function. The summary systems chart, which is attached to the paper, does not explicitly show the control mechanism, but the text stresses its existence and its method of operation.

Given the process oriented emphasis of the paper, the general management activities (i.e. those which do not specifically relate to the monitoring of particular functions), are not specifically discussed. Such management activities include, among others, financial management, personnel management, training, and generally activities related to the maintenance of a suitable "capacity".

Superimposed over the functional oriented structure, there must also exist an overall integration-co-ordination function. Only some aspects of such a function can be described formally within the present approach -- those with specifiable inputs and outputs. However, given its all-embracing nature, considerations relating to integration are understood to permeate the "ideal" statistical system and most of its component functions. Emphasis is given in the paper to more concrete aspects of integration; other aspects of it are emphasized in a separate section.

In reading the following sections, it is important to remember that our aim is to describe the functions of the Statistical System and not its organization. As far as possible, we have attempted to avoid the use of terms that have strong organizational connotations. However, where this was not possible, the reader must interpret such terms as describing functions and not as identifying organizational units. There is no implication that any particular subfunction is (or even should be) the sole responsibility of any particular organizational unit -- on the contrary, we believe that many of the subfunctions can only be carried out successfully by interdisciplinary teams. Conversely, any one organizational unit will in general have an essential role to play in several of the subfunctions identified here.

The difference between a line organization and the functional view presented
in this paper is fundamental in understanding our approach. Understanding
that functions and processes are quite distinct from organization facilitates
the identification of needs for mechanisms and procedures to ensure that the
institution, collectively, carries out all necessary functions.

THE OVERALL SYSTEM

In this paper, the purpose of the overall "Statistical System" has been
assumed to be: to provide to the public (including all levels of government)
coherent, relevant, timely, well understood and readily accessible statistical information on economic and social structures, processes and attitudes.

The term statistical information is meant to encompass statistical data (numbers), information about the reliability of such data, documentation of the
process used in compiling data, commentary and analysis*.

As we mentioned in the introduction, this system is not necessarily seen as
existing solely within the confines of a single agency, but rather as encompassing the provision of all statistical information by all involved government agencies, and for which there exists a common mechanism for medium-term
planning, and for the establishment of common standards, operating tools and
practices.

A degree of arbitrariness in defining the scope of the present paper is unavoidable. We have defined the overall system to encompass all of the explicitly statistical production processes, their planning, co-ordination and
priority setting, analysis, and the interactions between them. It does not
include what might be called the external environment of the statistical
service, except to the extent this may directly affect its internal functioning: the public from which the information is collected, users for whom the
information is collected, or administrative data bases which may be a source
of statistical information.

The methodology of the paper calls for the consideration of control for each
function or subfunction of the system - and, in fact, at this stage we should
consider the nature of control of the overall system. As mentioned earlier,
we use the term "control" as that activity which is designed to ensure that a
particular process carries out its assigned function in harmony with the rest
of the system. This concept, applied to the overall statistical system, implies an activity designed to harmonize the statistical system with its environment (public attitudes towards statistics and statistical data collection, social, economic, political and legal environment). Such harmoniza-
tion consists of the receiving of feedback from the statistical system as
well as its environment and the provision of stimuli designed to influence
the statistical system or, indeed, its environment.

---

*We will by-pass the issue of "how much analysis" and the distinction between
specifically policy-oriented analysis, versus statistical analysis high-
lighting in a "neutral" fashion "what the data show".

The most significant stimuli designed to influence the statistical system involve broad policy decisions on priorities which at lower levels of control
get translated into resource allocations. Each of the desirable attributes
of statistical information listed above (coherence, relevance, timeliness,
being well understood and readily accessible) imply shades of priority to be
assigned to different functions and subfunctions of the statistical system.
To the extent that priorities (choices) are involved, the choices are designed to steer the statistical system into a state of closer harmony with
its environment. At such a broad overall level the priorities are unlikely
to relate to particular projects but may well relate to broad programmes - and
thus this overall control function would influence the medium-term planning
function of the system. Other non-programme types of priorities are likely to
relate to functions: such as to strengthen the coherence of statistics (with
possible implications for new programmes,strengthening the analytical function
or its feedback to current operations, achieving better operational integration); or to improve the accessibility of statistics (with implications on
the dissemination function, on the function of maintaining appropriate data
bases and access methods); to improve the safeguarding of confidentiality of
identifiable returns; to reduce response burden; to attach measures of reliability to key series, etc.). It is customary to think of resource allocation
at the level of competing statistical programs or projects. The broad policy
type priorities indicated above will, of course, effect the resources to be
allocated to particular projects, or in fact the distribution of resources
to different aspects (functions) of particular projects - but this takes place
at a lower level of control. Finally, the control mechanism must approve the
medium term plan of the statistical system.

If the control activity is to carry out its harmonizing functions, it must be
in a position to obtain signals not only from the statistical system itself,
but also from its environment. In fact, it must visibly monitor the public
status, image and acceptance of the statistical system (public concern about
privacy and confidentiality, response burden and its distribution, the assessment of users' views concerning the coherence, relevance, clarity and accessibility of statistics, changing legal frameworks, etc). It may receive such
signals through the statistical system itself (complaints from respondents,
monitoring refusal rates, results of user conferences), or directly from outside the system (monitoring the media, through organized public or semi-public
hearings, etc.). Moreover, while its main feedback is to the statistical
system, inducing it to change in the direction of closer harmony with its
environment, the control activity will also attempt to change the environment:
by initiating requests for changing relevant legislation, by influencing the
interfaces between data collected for administrative and statistical purposes,
by publicly defending, if necessary, the statistical system, by influencing
government budgetary priorities, and through attempting to improve the statistical knowledge and awareness of the general public.

In some decentralized systems the so-called central statistical offices carry
out a large part of this control function; in others it is attached to other
ministries; in yet some other countries special legislative committees or
hearings carry out at least some of these activities. Some countries with a
central statistical system have established public bodies, like a board of
directors, to carry out some of these control functions.

It follows from the previous paragraph, as well as from well known principles
of cybernetics, that the type of control function discussed above must cons-
sciously be designed to receive signals from, and provide feeback stimuli to

both the statistical system and its environment. To this extent at least the control mechanism must be seen to maintain a degree of independent authority.

The outputs of the statistical system as a whole consist of statistical information as defined above. Its inputs include the priorities (both current and predicted) of governments and other decision-makers, models of how society and the economy behaves, specific requests for statistical information, data collected from respondents, and data obtained from administrative records.

The next sections deal with a detailed examination of the two main functions including, in turn, identification of their subfunctions, inputs, outputs, possible measures of effectiveness and the control measures which can be taken should the system not meet its criteria for effectiveness. Wherever possible, examples are indicated of skills (disciplines) required to perform a function or subfunction. It should be remembered that these are examples only, not exhaustive lists and also that the organization which might be implied by the examples is not necessarily being suggested as ideal. At the end of the paper a schematic diagram is provided which identifies the main functions, their subfunctions and some of the most important interactions and feedbacks among them. The reader is asked to refer to the chart as he proceeds with the paper. Circles are illustrative of tools used by each of the functions.

THE TWO MAIN FUNCTIONS

The overall system breaks down naturally into two main functions. The first (the top half of the attached chart) is the one of providing statistical information (that is, the obtaining, processing, and dissemination operations). The second, often less developed function (the bottom half of the chart), is that of maintaining and adjusting the framework within which the first one operates: medium-term plans, standard policies, tools, classifications, etc. It is largely through the second function that long term requirements, including integration requirements and changes which ultimately affect the "environment" of the first, are taken into account.

The first main function, referred to as "function 1", is defined to be:

> to provide statistical information to decision-makers and the general public. It is responsible for providing this statistical information as efficiently as possible within the existing framework as defined by "function 2" in the form of a medium term plan, appropriate operating policies, standard tools and classifications.

The main output of function 1 is, as mentioned above, statistical information broadly defined as above. Moreover, function 1 also provides information to function 2 about the nature and frequency of unsatisfied users' needs, data gaps and integration problems as revealed through internal analysis of data, missing policies, tools, concepts, classifications. Its inputs include short term priorities of governments and other decision-makers (longer term priorities being input to function 2), specific requests for statistical information, data collected from respondents or from administrative files, and the medium term plan, policies and standard concepts, classifications and operating tools established by function 2.

The components of function 1 are to:

(1.1) analyze requirements;

(1.2) identify the feasibility, priority and methodology of all other activities;

(1.3) assemble data,

(1.4) aggregate, analyze and interpret;

(1.5) disseminate;

(1.6) maintain the necessary data bases (including information about data, i.e. its reliability, documentation of methodology, etc.).

All of these components are considered to be separate subfunctions of function 1 and will be considered in more detail.

The effectiveness and/or efficiency of function 1 can be assessed by examining, for example, the use being made of statistical information being produced, the timeliness and cost of work done, the number and priority of satisfied and dissatisfied users, the amount of information obtained per "unit of respondent burden", and the extent to which the plans, policies and standards produced by function 2 are adhered to. Some of the major activities of the control mechanism of this function include adjusting project priorities, by allocation of resources between the components, by taking decisions on the accuracy of measurements required, particularly as they relate to integration and coordination objectives, and by adjusting charges to external users of what is produced. As will be seen later, the control activity can be largely identified with subfunction(1.2) above.

In summary, function 1 is the information producing and disseminating activity to which most statistical agencies are devoting the overwhelming proportion of their resources. In countries with a decentralized statistical system, function 1 may well operate within the confines of each of the different statistical agencies. In this case the control activity may well be vested in the head of each of the agencies involved. In centralized statistical systems the control activity of function 1 may be delegated to senior officers or subcommittees of the executive committee.

Function 2, as mentioned above, is designed

> to maintain the framework for the short term operating and analytical activities of function 1.

It is responsible, therefore, for establishing and obtaining the framework of function 1 in light of its projection of future requirements of decision-makers, and of longer term government priorities. Given the time lag between decisions relating to the need for certain types of statistics and their availability in a useful form (often time series), knowledge or prediction of longer-term statistical needs is an essential activity of function 2.

The inputs are: longer-term government plans and expected social and economic developments which might need statistical support or monitoring information; and signals from function 1 concerning the nature and frequency of unsatisfied user needs, reliability and integrability of current statistics, operating problems and concerns indicating the absence of critical statistical policies, standard tools, concepts, classifications.

The outputs consist of the creation and updating of proposed medium-term plans, new policies, standard concepts, classifications, tools and practices for the use by function 1. The medium-term plan is the vehicle through which longer-term statistical requirements are expected to be met, including projected needs, accommodating current unsatisfied needs, problems of integration and reliability.

The components of this function, each of which are to be discussed subsequently in greater detail, include:

(2.1) medium-term planning;

(2.2) development and promulgation of standard concepts and classifications;

(2.3) development and promulgation of standard tools and practices.

The effectiveness of this function can be measured primarily through the relevance of the information produced by function 1, i.e. the extent to which statistical information is available to decision-makers and the general public of the quality, format and timeliness required; the efficiency of the statistical system (e.g. lack of duplication of development in function 1 due to the absence of effective standard tools and practices); and the level of integrity of the overall system (the maintenance of appropriate standards of confidentiality, the overall level and distribution of response burden).

The control of function 2 itself (which is not to be confused with the control of the total statistical system) consists basically of the allocation of resources between the subfunctions. Since function 2 is, of necessity, organizationally diffused in most countries and statistical offices, the location of the overall control for this function should be at the centre.

It has been mentioned in connection with function 1 that it may take place, in countries with a decentralized statistical system, within several different agencies. By contrast, even though it may be organizationally diffused, the existence of a single function 2 is central to the notion of a single national statistical system.

FUNCTION 1 -- DETAILS

This is the "information-producing" function. At one end the set of requirements expressed by decision-makers and short term government priorities* come in, and at the other statistical information (as defined in Section 2) comes out. To get from one end to the other, six separate functions have been identified.

(1.1) Analyse Requirements

(1.2) Identify Feasibility, Priority and Method

(1.3)   Assemble Data

(1.4)   Analyse, Interpret, Transform Data

(1.5)   Disseminate Data

(1.6)   Maintain Meta and Micro Data Bases.**

They are related in the following fashion:

(1.1)   Analyse Requirements

This is the subfunction which, in over-simplified terms, can be thought of as receiving the signals from the outside world indicating demands for statistics, interpreting these signals and passing them on to the remainder of the system.

Both the notions of "receiving" and "interpreting" signals require some elaboration.

In reality, demands for statistics arise in a variety of forms and through a variety of channels. The detection of external demand is accomplished through formal means (conferences, committees, panels), ad hoc means (requests for certain types of statistical information), and informal means (personal contacts, attendance at conferences sponsored by academic and private research organization, etc.). Moreover, the demand may relate to the retrieval of tabulation from existing data bases, to the provision of new information, or to the need to change existing statistical information (it may be in need of improved reliability, more detail, better timeliness, increased frequency, changed concepts, etc.). Indeed, most subject matter replies within statistical offices correctly consider it to be one of their major functions to keep attuned to the external demand for statistical information.

Thus significant and varied resources, involving senior personnel, are devoted in most statistical offices to the detection of external demand which, oversimplified, is represented in our paper as an external signal. It is, however, at least conceptually, essential to separate the function of detecting demand from the several other functions which may be involved in satisfying it - otherwise it is far too easy to think in terms of meeting each new demand with new surveys.

----

* By short term is meant specific needs which may take 1-3 years to meet but for which the requirements are present now.

** Meta data is defined to be information about the data: commentary, description of concepts and methodology used, estimates of reliability, description of structure and form of storage, and method of retrieval. Micro data is the unaggregated data, including derived data.

In "interpreting the signals", clearly the first question is: can the demand
be satisfied from the existing data base? In order to answer even such a
relatively simple question, the particular demand must typically be translated: one can only search the existing data base in terms of the concepts
already utilized in the collection and storage of past data. Thus, ideally,
in order to carry out this function systematically, there should exist a
dictionary of standard concepts, the particular demand should be translated
into those terms (or the absence of a suitable standard concept should be
noted), directories should be available to indicate the particular data bases
containing information relevant to the concepts in question, and meta data
should be available to further describe the information which is available
(measures of its reliability, frequency of collection, amount of related
classificatory detail available, etc.). It is only at the end of the search
process implied by the above description that the question can be answered:
can the particular demand be satisfied from the existing data base? If the
answer is affirmative, a request is passed to the dissemination subfunction
(1.5) for action.

If the demand for information cannot be satisfied fully from the existing data
base, perhaps sufficiently closely related information exists which might
suffice for the purpose at hand. However, since there are no formal measures
of "distance" between concepts, the subfunction (1.1) must perform, working
together with the requestors, the intuitive task of identifying specific alternative standard concepts which might satisfy the need at hand and in terms
of which the existing data base may be searched.

If the demand cannot be satisfied, even approximately, from existing data,
then the potential exists for the generation of new data. The data generation
may or may not involve new collection as such, it may be restricted to the
generation of new micro or macro data through modelling and other analytical
techniques using as independent variables existing micro or macro data, including administrative records. And even if the demand can only be satisfied
through a new survey, it may well be decided that the demand should not be
satisfied at the present time. Whenever the demand can only be satisfied
through the generation of new data, the decisions concerning the whether and
how are not made in the subfunction (1.1). What this subfunction must do is
to articulate the demand into standard terms which would permit the next
subfunction, that of the identification of feasibility, priority and method
(1.2), to make an explicit decision about it. The standard terms would involve the concepts about which data are required, the outside priority attached to the data request, the implications for reliability of the intended
utilization of the data, amount of detail required, frequency, deadlines, etc.

The conduct of this subfunction is clearly a highly professional one. As
Goldberg (1) puts it, "The producer must base his evaluation of the demand on
fairly precise, even at times inferred or assumed, information about utilization". The penetration of the users' mind, whose need is generally articulated in vague terms, the "inference" and "assumption" about the intended
utilization is the essence of this function. At the same time, one must
emphasize that the decision about the disposition of requests for data should
be made in subfunction (1.2) - except when the request can be satisfied from
existing data. The reason for this emphasis is a basic frugality of approach
which we want emphasized. The statistical system must have mechanisms whereby, on the one hand, it ensures the most extensive utilization of existing
data, and on the other hand it avoids sliding imperceptibly from a request
for data to a new survey. This can only be achieved, we believe, through a

formal separation (not necessarily in organizational terms) of the function
of "receiving signals" from those of responding to them. However, for the
rest of the system to be able to respond to the largely unstructured signals,
they must be translated into what has been called above "standard terms".

The requirements examined in this fashion can be either new or old -- it is
here that the task is carried out of ascertaining whether the requirement for
information supplied on a regular basis still exists, or whether changing
requirements for data or its reliability should signal a need to review on-
going projects. Function (1.1) does this partly on the basis of the regular
(formal and informal) contacts with the users and through monitoring summary
statistics or dissemination.

The inputs of this subfunction are, thus, the various expressed needs for
statistics, short term government priorities, summary information about the
dissemination programme involving currently produced statistics, standard
concepts and classifications provided to this subfunction by function 2, and
meta data about the currently available statistical data base. The outputs
consists of requirements translated into standard terms; this output is
either directed to the dissemination subfunction (1.5) if it can be satisfied
from the existing data base, or to the subfunction (1.2) where feasibility,
priority and methods are determined. There is one exception to this, i.e.
if the data requirement is clearly unfeasible (e.g. it would contravene existing legislation or blatantly exceed the internal capacity) - in which case it
is rejected at this stage. Another special type of output consists of feedback to function 2 concerning missing standard concepts and definitions.

The control of this subfunction examines the proportion of "misdirected requests" and the turnaround and cost of (1.1), and then adjusts resources
within (1.1) or provides feedback to the second system about missing concepts
and classifications.

At the present time only on rare occasions do the requirements appear to be
stated explicitly before moving on to function (1.2). All too often, particularly if the information does not appear to be readily available, the requirement is expressed in the form of a need for a survey of some size, rather
than for statistical information with some specified accuracy. Yet, without
a degree of such specificity, the feasibility study component of the next
function (1.2) becomes largely meaningless, and almost impossible to evaluate.

(1.2)   Identify Feasibility, Priority and Method

This function receives requests from (1.1) (analyse requirements), couched as
far as possible in standard terms. These requests concern data needs which
do not appear to be satisfiable through available, releasable information.
It also receives signals from the analytical function (1.4) with respect to
gaps and inconsistencies detected in the data base and derived from the confrontation of different data within the frameworks provided by social and
economic models.

Function (1.2) has to assess the request from several points of view. First
of all, it has to assign a preliminary priority to it. This priority is
based on a number of considerations (1): the auspices of demand (i.e. demands
have a greater chance of being satisfied if they represent the general national interest); the extent to which the statistics would satisfy a variety of

needs; the short-run internal capacity of the statistical system; and last
but not least, the extent to which the satisfaction of the request is in line
with the objectives of the medium term plan, including integration objectives
related to filling certain data gaps in economic and social models. Particularly if the data requirement does not fit into the medium term plan, an
important prerequisite for its satisfaction is likely to be the availability
of external funding.

The assessment may result in the rejection of the request, or its further
pursuit. In the latter case this subfunction specifies alternative feasible
methods, estimates the resource requirements of each and recommends a strategy. The strategy might involve the use of releasable information*, the
assembly of data requiring collection, or the use of available data together
with modelling techniques, analysis and interpretation.

The recommended strategy is discussed with the requestor and, assuming the
requestor's approval is obtained, a directive is sent to the appropriate
subsequent function for action. Should no feasible strategy be found which is
acceptable to the requestor, the request is rejected. In practice, of course,
these "rejected" requests are often shelved until sufficient pressure is
brought to bear so that the system accommodates the request either under con-
tract or as part of its on-going workload. However, one can conceptually regard these requests as being rejected and fed back into the system later.

The requests are not necessarily handled one at a time. One can imagine several related requests being considered in conjunction with one another: e.g.
the collection of a variety of information in the form of supplementary questions added to an on-going household survey program. And at any rate, all
requests are evaluated always in terms of the broad overall criteria listed
at the beginning of this section which, together, tend to ensure a degree of
short-run coordination in the satisfaction of demands (longer-run integration
and coordination objectives are incorporated into the medium-term plan developed by function 2).

This subfunction must clearly have access to information (project accounting
information) about the cost of the subsequent subfunctions, and have responsibility for applying the first system's priorities and scheduling the work of
these other subfunctions. It should be noted that formally (in terms of
visible inputs and outputs) three other functions, (1.3), (1.4), (1.5) (collect, analyze, interpret and disseminate) communicate only through this one.
Thus a directive to collect data incorporates, implicitly, a directive to
analyze and interpret after collection and then to disseminate. In a very
real sense, therefore, this subfunction coincides with the overall programme
control activity of function 1.

---

*Even though (1.1) determined that releasable information does not meet the
requirement, a reassessment may well show that available information provides
a suitable "proxy" for the particular need.

We see this function being carried out at a senior management level of statistical offices, assisted by interdisciplinary teams put together for specific projects or groups of projects from different organizational units for the duration of the study only. The sort of skills required are those which would normally be found in a senior project team -- including subject matter, survey methodology, systems, data processing, operations, field collection if appropriate, etc.

The effectiveness of this subfunction can be assessed in terms of the proportion of misdirected requests, turnaround and cost, the actual time, cost and reliability achieved in the other subfunctions relative to the estimated ones used in the feasibility study, the level of respondent dissatisfaction and the amount of unnecessary overlap between surveys.

We can summarize the inputs of this function as consisting of information about all data available in the statistical system, requests for new information expressed in standard terms by function (1.1) (analyze requirements), the medium-term plan prepared by function 2, results of market studies and statistics on data uses from the dissemination function (1.5), assessments of reliability of on-going surveys from the data assembly (collecting) function (1.3), perceived gaps or lack of coordination in the data base from the analytical function (1.4), and accounting data. The outputs of the function include directives to (1.3) to assemble data requiring collection, to (1.4) to aggregate, analyze or transform data, to (1.5) to disseminate data, rejected requests, and feedback to function 2 indicating unsatisfied data needs to be considered for inclusion in a revised medium-term plan, and missing policies, concepts and tools. Within our "ideal statistical system", whenever the output includes a directive to (1.3) to assemble data requiring collection, the same type of specification must be present as discussed in connection with the previous function -- except that it is now refined through a feasibility study.

(1.3)   Assemble Data Requiring Collection

This is the subfunction within the first system which has direct contact with the sources of data -- either potential respondents to a questionnaire, or administrative records -- as well as with the micro and meta data bases of the statistical system. It receives its directives to assemble data from function (1.2).

Involved in this function are the preparation and testing of the detailed design of surveys, manipulation of administrative files (e.g. record linkage), or a coordination of the two. In the case of surveys this would include: determining the survey frame and the sampling method; the questionnaire (when applicable); the survey methodology (i.e. the collection and processing methods); as well as the survey operations (i.e. the actual collection, processing and evaluation operations). Also included here are the maintenance of tools of operational integration, e.g. appropriate registers, maintenance of multi-purpose survey vehicles - thus like all other subfunctions, this function also makes its contribution to integration.

Given a certain response load, determined by the totality of surveys approved in (1.2), the equitable distribution of response burden is also part of this subfunction. Finally, the maintenance of a certain level of operational capacity (qualified manpower, technology, operating procedures), in line with the requirements of the medium-term plan, is also the responsibility of (1.2).

There is a natural checkpoint after the details are settled and before collection begins, and another (which includes quality assurance) after collection and processing are complete but before meta and micro data are added to the bases via the data base maintenance function (1.6). These checks are part of the function itself and are typically performed by managers of participating areas.

It is important to draw a distinction between the aspects of determining the collection methodology carried out in functions (1.2) and (1.3) respectively. In (1.2) the feasibility of very broad alternative approaches is considered -- e.g. is collection necessary or can the required information be deduced from available data; can record linkage between different files yield the required information; what types of survey frame are available; can the collection be carried out as a supplement to some existing survey or is a new survey required; etc. Given the broad approach, the detailed design and attempts to optimize it are part of (1.3) -- as well as the conduct of the corresponding operations.

The main inputs to this function are the following: directives from function (1.2) to assemble data, information collected from respondents, administrative records (both as sources of statistical information and updates to registers), data from previously collected surveys and meta data, (e.g. cost data, reliability data concerning alternative designs, evaluation of the effectiveness of different operations, etc.) and the standard tools, practices, concepts and classifications provided by function 2. Examples of standard tools used here are policies concerning the use of registers for surveys, generalized sample selection, edit, imputation systems and tabulation systems, etc.

The outputs of this function include, first and foremost, the micro data resulting from the collection operation. Certainly the micro data output should include clean microdata (i.e. edited, imputed, weighted) and, depending on possible future needs, unimputed data as well. It should also include what we call meta data, i.e. all relevant information about the data: definition of concepts used, reference period, measures of reliability, details of methodology, description of structure and form of the data stored, methods of retrieval. The output also includes information on costs, turnaround and, for on-going surveys, the results of quality monitoring which may include the possible need for a redesign.

The effectiveness of this function is measured in terms of the turn-around, cost and quality of the data assembly function, and level of respondents' dissatisfaction.

The skills required for performing this function include, depending on the particular project, project management, subject matter, survey design, systems design, field collection, various other operational skills.

(1.4) Analyze, Interpret, Transform Data

This function operates on the micro and meta data bases (including administrative files) and, at least formally acts on directives from function (1.2).

We say formally, because in our "ideal" system a directive to undertake certain routine analysis is understood to accompany each directive to assemble
new data; thus an on-going analytical activity is understood to accompany any
on-going data assembly activity. Nevertheless, the level and disposition of
the total analytic capacity must be under the control of the program control
function (1.2). There are, of course, many different kinds of analytical
activities which are all included under this overall heading: interpretation
of data arising from a single or several different collection activities,
preparation of highlights for publications, evaluation of the extent of co-
ordination and integration through analysis and confrontation of data within
economic and social models and accounts, preparation of indices and aggregates
(e.g. the National Accounts), transformation of data through estimation,
modelling, projection, etc.

In many formal respects, this function is analogous to (1.3). Whenever there
is a new analytical method to be used, it is prepared and tested, rather like
in the preparation and testing of the detailed survey design. There are analogous checkpoints in this function to those of the data assembly function
(1.3), after the preparation and testing, but before implementation of the
method, and again after evaluation before new data or meta data are sent to
the micro and meta data bases via the data base maintenance function (1.6).
There is a more important analogy as well: the output may consist of new
data, in addition to the more customary analytical material. Such new data
can be in the form of aggregates (e.g. economic accounts, input-output tables), indices (e.g. price or employment indices), modelled disaggregation
(e.g. synthetic small area estimates), and derived micro data (e.g. the results of synthetic linkage of micro data between surveys, or income or production estimates derived at the micro level from taxation data).

Again, like (1.3), the control is exercised by examining the turnaround, cost
and quality of work carried out and a corrective action is taken accordingly.

The skills we see being used here include the different types of subject
matter analyses, econometrics, those used by people who work on the System of
National Accounts or System of Social and Demographic Statistics, and statistical methodology.

The formal inputs to this function consist of directives from (1.2) and the
available micro and meta data bases. Of course, close contact with the collection function (to maintain an awareness of the implications of collection
methods) and with key users is part of the natural mode of operation of this
function.

The outputs are analyses, commentary, description of concepts and methodology used, estimates of reliability of the new data produced here, new aggregate
data, new micro data, and information on costs. A specially important type
of meta data produced here involves assessments of the extent of statistical
coordination and integration of the data bases and its feedback to both subfunction (1.2) for shorter term corrective action, and to function 2 for the
planning of more fundamental corrective action through the medium term plan.

(1.5) Disseminate Data

The raison d'etre of the statistical system is, of course, its output. It is partly for this reason that we broke this subfunction out as a separate one. In a more traditional view dissemination would have been grouped with the largely operational subfunction of data assembly (1.3), one of its last activities being the production of a publication. We want to emphasize that dissemination is a far broader activity than the output of publication. It emcompasses the output of statistical information in a variety of formats and using different media, in order to maximize the usefulness of data for its users. Thus, in addition to the production of publications, the production of aggregate statistical information on summary tapes, microfilm, microfiche is also included here. Further, to the extent that a databank of statisical aggregates is maintained (either in the form of a library of printed publications, or in the form of a machine readable data bank), this too is part of (1.5). The maintenance of such a "library of releasable statistical information" (as it is shown on the attached chart) involves the storage of aggregate data on the appropriate media, i.e. the choice, out of the millions of potential data points, of those which should be stored, and the choice of the medium for each data point on which it should be stored. Difficult market forecasts are involved in such choices, with significant cost/benefit implications: failure to store an aggregate implies that, should it be needed, it can only be reproduced as special tabulation from the micro data base at relatively high cost; storage on paper or microfilm/microfiche is the next cheapest alternative, but it is not nearly as accessible as data stored in machine readable form within a data bank.

Since this subfunction is responsible for the production of all aggregate data from the statistical system, it is clearly responsible for the application of appropriate confidentiality protection routines - supplied to it in the form of policies or algorithms by function 2.

One of the major motives of the present paper is to stress the mechanisms required within the statistical system to foster the extensive and intensive exploitation of the accumulated data capital. One important means to achieve this is the encouragement of what Goldberg (1) calls secondary utilization of data, i.e. utilization which goes beyond the satisfaction of those initial statistical needs which represented the initial justification for the collection of a particular type of survey data. This marketing activity is also a part of the dissemination subfunction, together with more formal market studies of the needs of the most important user groups.

Finally the satisfaction of ad hoc tabulation needs (i.e. the meeting of demands which can be satisfied from the micro data base of the statistical system, but not from the library of releasable statistical information) is also part of this subfunction.

Formally, this function acts on directives received either from the analysis of requirements subfunction (1.1) - if the requirement can be satisfied from available micro or aggregate data, or from the program control function (1.2) if the dissemination involves new surveys or otherwise newly created micro data. In practice, some of the requests for data bypass function (1.1), particulary in the case of regular subscribers to published data, or of users with access to publicly available data bank information: the translation of user needs into standard terms would in such cases either have taken place earlier, or might have been carried out by the users themselves. At any rate, however, this function provides summary information on data use to function (1.1) and market study information to (1.2).

Its <u>control</u> examines the cost and turnaround of access to the "library", the
extent of public awareness of the availability and uses of statistical information, the effectiveness of producing ad hoc tabulations, extent of secondary data utilization, the relevance of market information, and then takes
appropriate corrective action.

<u>Inputs</u> consist of: directives from (1.1) or (1.2) to disseminate, micro and
meta data, the library of releasable information, laws, policies and standards relating to data dissemination.

The main <u>output</u> is, of course, stastistical information (broadly interpreted,
as outlined in previous sections), accounting data and the results of user
liaison and market studies.

Examples of this operation are found in most statistical offices, at least in
an embryonic form. They include the various publication distribution, dissemination, use development and user liaison functions.

(1.6) Maintain Meta and Micro Data Bases

Looking at the attached chart, or reading the previous sections of the paper,
the central importance of what is referred to as meta and micro data bases is
quite apparent. The particular collection or analytical activity which is
depositing data in the bases is by no means the only one having access to it.
Thus the notion of a corporate data base assumes a paramount importance in
the context of the statistical office. This is what Nordbotten called the
data capital of the statistical system.

Note that, implicitly, the entire functional presentation of this paper fundamentally hinges on this function: (1.1) could not search the data base attempting to find information relevant to a given request, (1.5) could be
severely handicapped in its dissemination and marketing activity without it,
the scope of (1.4) to analyze data arising from different collection sources
would be severely limited without this function.

The safeguarding of the data capital on behalf of the entire statistical system and all its present and future users is the heart of this function. The
safeguarding consists of maintaining the integrity of the data base i.e. its
physical security, as well as ensuring that the data base and its documentation are consistent (e.g. that the data would pass the edits indicated in the
documentation); the maintenance of appropriate documentation sufficient to
indicate the content of the data base as well as its physical structure.
This subfunction also maintains suitable access mechanisms (including software) to facilitate retrievals from the micro and meta base. A particularly
important activity of (1.6) is the maintenance of data directories, i.e.
directories showing what physical files contain information about particular
standard concepts.

Just like the description of a given function does not imply its separate
existence in the form of a line organizational unit, similarly the fact that
meta and micro data bases are shown on the attached chart as a separate entity
does not imply the existence of a single, physically separate data base. It
does imply, however, the existence of the subfunction (1.6) - probably carried
out within a variety of organizational units, but carried out subject to
common policies and standards of storage and documentation and in such a

fashion as to facilitate broad access to the totality of the data bases within the statistical system. The utilization of standard concepts, incorporated into standard data dictionaries and directories, is a key to this function carrying out its mandate.

The control of this subfunction is carried out through the examination of the accessibility of micro and meta data, the turnaround and cost of data access, audits of data security and integrity. It can adjust policies concerning the organization of, or access to, the data bases and/or take other appropriate corrective actions if necessary.

Inputs to this subfunction are the micro and meta data from the data assembly subfunction (1.3) and from the analytical subfunction (1.4). Its output is the set of updated meta and micro data bases.

FUNCTION 2 -- DETAILS

This is the function which "maintains the framework" for the operation of function 1. The framework, for purposes of the present paper, consists of the medium term plan, and standard policies, tools, classifications and concepts. As mentioned in Section 3, whereas one can visualize several statistical agencies each carrying out all or most of the activities involved in function 1, the existence of a single national mechanism for carrying out the activities of function 2 is considered to be vital for the existence of an overall, co-ordinated national statistical system.

The inputs, outputs and measures of effectiveness of function 2 as a whole are described in Section 3.

The component subfunctions of function 2 are the following:

(2.1) Medium-Term Planning;

(2.2) Develop, Maintain, and Promulgate Standard Concepts and Classifications;

(2.3) Develop, Maintain, and Promulgate Standard Tools and Practices.

The details of these subfunctions are described below.

(2.1) Medium Term Planning

This subfunction plays a central role not only within function 2 but within the whole statistical system. It analyses and abstracts needs identified both inside and outside the statistical system and prepares a medium term plan which provides the operating mandate for function 1. In a very real sense this function forms an essential part of what might be termed the "corporate management" of the statistical system.

The medium-term plan, as the action program of the statistical system, has to incorporate a number of different aspects. Specifically, it has to provide for the following:

    a.   new statistical programmes or projects and the phasing out and/or replacement of lower priority projects

    b.   requirements of co-ordination and integration

    c.   functional and/or structural changes required

    d.   the determination of a level of capacity for function 1 which would permit it to respond to short-term needs not incorporated in the medium-term plan.

While the present paper is not a treatise on planning, a few words might be in order in connection with the four points above.

With respect to a., the inputs consist of the recorded unsatisfied needs from function 1, the results of market studies and monitoring the dissemination activity (also from function 1), longer term government and other user priorities, anticipatory analyses of likely future needs arising from new social, economic or government programme concerns (a particularly important activity given the time lag between perceived needs for statistics and the availability of usable statistical series), and perceived needs within the analytical and/or data assembly functions of data gaps or other weaknesses within the current statistical output of function 1. The sum total of these demands on the statistical system is typically greater than what can realistically be expected to be satisfied. The factors affecting the choice are well discussed in (1) and include: the social milieu (the level of support the statistical system can expect, both financially and in terms of public support to live with a level and type of response burden). The auspices of the demand (i.e. the extent to which a proposed new statistical project or programme is in support of broad national objectives); the extent to which the demand serves multi-purpose and integration objectives (i.e. the extent to which it would satisfy a variety of users, or whether the information in question is a significant element in a more comprehensive statistical aggregate); the extent to which it would fill important data gaps affecting the co-ordinated analysis of statistical series; the survey capacity of function 1 (i.e. the extent to which function 1, particularly subfunction 1.3 can be expected to cope with a given level of survey taking and processing, in terms of both human and capital resources); developments in the art and science of measurement.

With respect to b., it is important to emphasize that the medium - term plan is one of the major vehicles through which detected integration and coordination problems can be addressed. Problems of integration can, of course, be of different kinds: data gaps in the major aggregates on economic and social models - these, if sufficiently important, can receive priority consideration among the new statistical projects; lack of suitable standards - which could result in the need to develop or update the appropriate standards by subfunction (2.2) and corresponding revision projects in related surveys; lack of co-ordination of the coverage between surveys, particularly economic ones - this may lead to priority being given to the development of suitable control mechanisms, such as a single register of businesses, etc. While the number of potential integration problems and possible remedies for them is numerous, the main point is that they are most effectively handled within the mechanism of a medium-term plan.

Similarly, changes in the kinds of functional policy priorities, discussed in Section 2, can also most effectively be envisaged within the medium-term plan (e.g. improve the overall dissemination effort in some specific manner, strengthen the analytical capability within the statistical system, implement policies related to such concerns as the safeguarding of confidentiality or reduction or redistribution of response burden, attach measures of reliability to key statistical output, etc.). Such functional policy priorities may result in specific projects, which cut across survey boundaries, or they may affect the resource requirements (both in magnitude and kind) of individual statistical projects. Either way, since longer term resource commitments are involved, such policies and their implications should be considered within the framework of medium-term planning.

Finally, the question of internal capacity must also be considered. Clearly, the plan must not exceed the internal capacity of the statistical system, or at least the plan must be in line with a realistic assessment of the rate at which the internal capacity can be developed (or diminished). However, some "surplus" capacity must also be planned for: function 1, while it operates within a medium term plan, must still be capable to respond to short-term government (or other) priorities for new statistical information. Such short-term requirements may or may not be accompanied by fresh funds; however, in many respects the internal capacity of the statistical system can only be altered relatively slowly and thus the capacity may not be present to accommodate the short run requirements, even if fresh funds are offered. Without a determined effort within the medium-term plan to keep such a margin of "contingency" capacity, the short-run priorities often have the impact of disrupting the orderly implementation of the medium-term plan.

The point has been made repeatedly about all the subfunctions that they are not to be construed as organizational entities. This has to be reemphasized with respect to (2.1): the planning and analysis involved in this function must permeate the statistical system, it cannot be genuinely meaningful operating within an ivory tower. Thus, this function must have very close ties with function 1, as well as with the key users (including potential ones) of statistical information.

The executive officers of the statistical system must collectively carry out this corporate function, although they may well be assisted in the integration of plans by special secretariats, a body of "corporate directors", or other mechanisms. In the "ideal system" of this paper, planning is a function with a highly visible output, possibly achieved only after several iterations.

In summary, we can view the formal inputs and outputs of this subfunction as follows. Its inputs consist of the need for statistics as identified both inside and outside the statistical system, i.e. unsatisfied but persisting needs from function 1.2, longer term government and other user priorities, including anticipatory analyses of likely future needs arising from new social, economic or government programme concerns. An essential input involves access to the meta data generated by function 1 concerning data gaps and the reliability of currently produced statistics, classifications and standards used, including their shortcomings, missing tools and policies, the extent of response burden and public reaction to it. Its output consists of new medium-term plans, directives to (2.2) and (2.3) to develop, respectively, new concepts and classifications, and new tools and operating standards.

The measure of effectiveness of this function is the existence of a viable
body of policies and a visible medium-term plan which carries the approval of
the user community and the Treasury (both possible represented within the
control mechanism of statistical system, discussed in Section 2), which function 1 is committed to carry out, and which fits in with the total current
and future capacity of the statistical system.

(2.2) Develop, Maintain and Promulgate Standard Concepts and Classifications

The importance of this function is clearly implied by much of the paper. For
the sake of brevity, we will only present a summary of this subfunction. It
acts on directives received from (2.1) and develops, tests, evaluates, and
maintains standard concepts and classifications. A suitable checkpoint is
present and enforced before the final step of promulgation of these standards
for use by function 1. Its inputs are from function 1 in the form of meta
data concerning the limitations of existing standard concepts and classifications, and directives from (2.1) to develop new ones. Its output consists of
the revised and tested new standard concepts and classifications.

Its control examines the feasibility of implementation of the new standards
by function 1, their effectiveness (measured by user reaction), the turnaround, and the cost of development.

The existence of standard concepts and classifications is fundamental to the
framework of function 1 as outlined earlier. They provide a large part of
the operational framework for functions (1.1), (1.5) and (1.6) (analyse requirements, disseminate data, and maintain data bases) and the conceptual
framework for (1.3) and (1.4) (assemble data, and analyse and interpret).
They are also essential tools of integration.

(2.3) Develop, Maintain and Promulgate Standard Tools and Practices

The efficiency, effectiveness and the integrability of the output of function
1 is largely determined by the existence of proven standard tools and practices. Examples of standard tools include survey design strategies, i.e.
concrete methods designed to achieve near-optimal balance between different
components of survey errors; methods to achieve survey designs with measurable
errors; design methods to balance the requirements for national and smaller
area estimates; collection tools, such as standard interviewer training man-
uals; general processing systems such as sample selection, edit, imputation,
record linkage, or tabulation systems; proven methods of public relations;
policies with respect to the distribution of response burden; policies concerning methods whereby users are informed of the reliability of statistics
produced by function 1.

This function is clearly of a research and development type. However, because
its output consists of standard tools and practices, these must be thoroughly
tested and evaluated before their promulgation. Thus the function contains
a checkpoint after development, testing and evaluation, but prior to promulgation.

As with function (2.2), this function must also be thoroughly rooted in the
operation of function 1, particularly (1.2), (1.3) and (1.6).

Its effectiveness is evaluated in terms of the feasibility of implementation
of the tools, their contribution to the efficiency and effectiveness of the
operation of function 1, as well as their contribution to the methodological
integrity, substantive integration and accessibility of the data base produced by function 1.

COORDINATION AND INTEGRATION

The system orientation of the paper necessarily results in emphasizing concrete processes, with well defined inputs and outputs. As a consequence, it
is hoped that the vertical aspects of the statistical system, which link different functions more or less within a given project or group of related projects, are reasonably well articulated. At the same time, the authors are
very conscious of the fact that this method of presentation cannot properly
highlight all aspects of what S.A. Goldberg (2) calls the "horizontal dimension" of the statistical system.

Some direct quotes from the paper referenced above explain well what we have
in mind. "The horizontal dimension ... relates to across-the-board inter-
divisional and interbranch activities". "This dimension is less visible than
the vertical; it does not appear in organization charts, except perhaps in an
impressionistic way. But it is no less important, especially in a statistical
organization concerned with the integration of its outputs (that is, that the
myriad pieces should somehow fit together) and the co-ordination of its programmes (that is, that balance is maintained among the various projects and
activities)." "The day-to-day external pressures concentrate on getting
things done within the sections and divisions -- the meeting of deadlines
within specified periods of time and the like. The impact of the horizontal
dimension is rather more subtle -- it is, or should be, reflected in the
guiding philosophy of the organization; it should permeate the actions and
policies in all parts of the organization. It should provide a corporate
consciousness, maintain checks and balances in the face of differing pressures
from the individual sections, foster inter-disciplinary project planning and
execution and overcome barriers, real or imagined, between the various parts
of the organization. It should ensure that common concepts, definitions,
classifications and methods are not only available but actually implemented
in the various divisions and sections so that the statistical series represent elements of an integrated framework and are as consistent and comparable
as possible."

The same paper lists some tools used in rendering operational the horizontal
dimension: standard classification systems, business registers of enterprises
and establishments, central questionnaire control, devices for ensuring the
adherence to standards, the system of national accounts and balances, and,
where they exist, a central field organization, central programming and systems design. All of these tools have either been touched upon in our paper
explicitly or, at least, they naturally fit into it implicitly. Some additional aspects of the horizontal dimension can also be identified in our presentation: the survey design aspect of function (1.3), the marketing aspects
of the dissemination function (1.5), function (1.6) which maintains the micro
and meta data bases, and almost all of the activity carried out by function 2.
However, as the United Nations paper points out in connection with the National Accounts, "to carry forward the unifying influences of the accounts, special machinery must be established within the organization". The same applies

to the other tools and standards.

This machinery may take the form of inter-divisional task forces, panels, committees; in a decentralized system the central statistical office carries responsibility for this machinery; in a centralized system special staffs may be assigned to keep it alive.

The present paper is not the proper vehicle to elaborate this activity further. The tools and some of the mechanism which play a vital role in the "horizontal dimension" are outlined -- however, what does not come across, and can only be emphasized here, is that, notwithstanding the presence of tools, only the unwavering support of the head of the statistical system, his/her executive and managers can keep the horizontal influences alive. The larger the statistical system, the greater the danger for its fragmentation -- and by fragmentation we do not necessarily mean organization, but rather the fragmentation of the coherence of the total statistical information base of the country. Thus, even where this is not explicitly evident, we ask the reader to appreciate, partially superimposed and partially embedded in the chart attached, the existence of viable and fully supported horizontal channels and connections.

CONCLUDING REMARKS

As indicated at the outset, the purpose of the paper was to present a description of an "ideal" statistical system in functional terms, i.e. free of the necessary constraints and compromises imposed by organizational considerations. Having prepared such a "model", it is entirely desirable to return to considerations of implementation - which inevitably involve organizations.

As a conclusion to the paper and in order to facilitate discussion, we would therefore like to highlight, in the form of a series of questions, what we consider to be some of the most difficult problems involving any concrete implementation of the above model.

What is the appropriate mechanism which should exercise the overall control of the statistical system? It should clearly represent the views of key users (including, in a federal system, both federal and provincial users), should have sufficient stature to be able to influence the external environment of the statistical system and should operate within the realistically assessed financial constraints imposed by the government - particularly as it considers the medium-term plan.

Through what concrete means does the overall control mechanism exercise its influence over the statistical system (including both functions 1 and 2).

The control of function 1 is relatively more clear: it presumably rests with the top executive officers in a centralized statistical office and with heads of different agencies in a decentralized system. However, the control of function 2 is more difficult conceptually. In a decentralized system what are the appropriate mechanisms for the key personnel of function 1 to be effectively involved (as they should!) within function 2; and conversely, in a centralized system, how should one ensure that the daily pressures (usually emanating from function 1) do not squeeze out from appropriate consideration the key activities involved in function 2?

The subfunction of analysis of requirements (1.1) would typically be carried
out by a variety of individual subject matter statisticians. Clearly, they
would need to have some guidelines and training with respect to the key activity of "translating user requirements into standard terms". Are there
any country experiences with respect to such guidelines?

It has to be recognized that an external user cannot be expected to distinguish between officials of the statistical office wearing their different
functional hats. Yet, a clear distinction must be made between the analysis
of requirements and the committment of resources. The latter is the privilege
of subfunction (1.2) - which considers particular requests in the light of a
series of broader considerations outlined in the paper. But what is the most
effective method of operation for the program control subfunction (1.2)?

A key tool for the effective operation of the analysis of requirements (1.),
the data assembly (1.3), the maintenance of data bases (1.6), and the dissemination (1.5) subfunctions is the existence of standard concepts, embodied
into a data dictionary. We would certainly welcome a discussion of the methodology employed by countries with experience in this field with respect to
the creation and maintenance of such data dictionaries.

The data dissemination subfunction is the primary channel through which statistical information is communicated to users and marketing information is
obtained from them. Even though the full activity implied by this function
may not be organized as a district operating entity, if it is to operate
effectively from the users' point of view, overall functional guidance and
policies are needed. What are the different country experiences and plans in
this regard?

A similar question to that raised in connection with data dissemination is
equally relevant with respect to the data base maintenance subfunction.
Through what technical and policy means is the "corporate data base" and access to it maintained? Is there a useful distinction between the global concept of a corporate data base (containing the most "important" data items)
and the much broader concept of "all data items ever collected by the Statistical System"? If the answer is in the affirmative, what data items should
qualify for inclusion in the "corporate data base"?

Are there country experiences with respect to a systematic method of storing
meta data?

What are the guidelines in different countries, which have a medium term plan,
for the maintenance of a margin of capacity to satisfy short-term requirements
not included in the plan?

(1) The organization of national statistical services; United Nations Economic and Social Council, Statistical Commission, Document E/CN.3/495, June
1976.

(2) S.A. Goldberg: The demand for official statistics and their utilization
in Canada, with special reference to the role of the National Accounts; Bul-
letin of the International Statistical Institute, 36th Session, Volume 42,
Book 2, Sydney 1967.

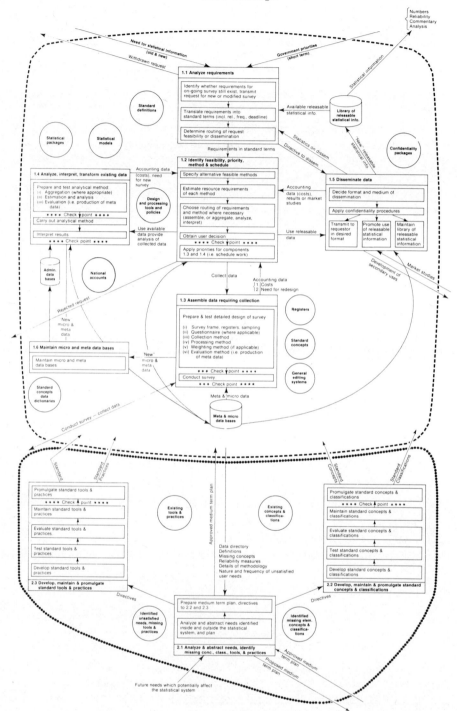

Functional chart of the statistical system

Discussion

The systems approach to statistical services as presented in the Canadian
paper "Functional Analysis of an 'Ideal' Statistical System" was very helpful
to each country as it provided a systems view of the functions and interrelationships that the seminar had been discussing. It presented a view of
the parts and functions of the statistical system which is applicable to both
market and centrally-planned economy countries. It will be useful in identifying gaps in the present system and as a checklist to make sure that central
statistical offices are in fact performing all their functions. Statisticians
need to follow the rules of science; they need a complete systems approach.

The discussion of the roles and functions in an ideal statistical system dealt
primarily with two questions:

1. What is the role of central statistical offices with regard to analysis?
How far should they go in analysis?

2. Will the system work?

Analytical Role of Statisticians

There are five stages in the production of data: data file development,
tables development, description, analysis and political interpretation. The
participants agreed that the determination of data files and tables, although
this could be interpreted as political and is part of analysis, is a task
that central statistical offices cannot avoid if they are going to perform
their functions. As to how far central statistical offices should go in stage
4, analysis, is a question that each central statistical office will have to
decide for itself. They will have to weigh, on the hand, the advantages of
giving a full analysis and,on the other hand, the risk of being accused of
not being objective. The choice will vary from country to country.

Sir Claus Moser suggested that the distinction should be made between policy
and politics. It is necessary for the statistician if he is doing a good job
to think about what he is doing; this is analysis. Much analysis is inevitable and there is a risk. To protect their image he suggested central statistical offices try to achieve the goals for protecting the integrity of the
statistical system outlined in the UN paper on organization, and strive to be
objective while realizing that the sceptics, may doubt their objectives. As
an example of the type of analysis he was recommending, he described a report
developed by a group of U.K. economists which meets monthly. Their report not
only gives the main economic indicators and new statistics, but also gives two
or three pages of analysis and interpretations. It does not make forecasts
or political conclusions. The Government's chief economic advisor, who is a
political figure, provides on top of the report one page of conclusions which

he draws from the report.   The U.K. is also now providing a similar social
report.

In "analysis", what is hoped for is the illumination of underlying trends.
To do this, closeness and participation in model-building is helpful.   The
degree to which central statistical offices participate in the design and use
of models varied among the countries.   In the U.K., the Treasury Department
does the main model building with the central statistical office providing the
data input for the model.   The central statistical office is responsible
neither for the policy extensions that go into the model nor the policy interpretations.   Norway and Denmark are responsible for the development and up-
keep of their countries' econometric models, but the use of the models is
left to outside agencies.   These agencies determine the input to the models
and they interpret the output.   More involvement in model-building is evi-
denced by Statistics Canada, which does a substantial amount of modeling because they hold the micro data to do the modeling and because they use the
models to test the fundamental quality of the data they are producing.   They
compare the figures given in one of their projection models with those computed in their national accounts to find "interesting figures."

This discussion of the analytical role of statistical offices was greatly enhanced by a discussion of involving more substantive participation by statisticians.   Chief statistical officers should make certain that the statisticians are aware of the policy concerns and are knowledgeable about the social
and economic policies so that the data collection will be effective in meeting
the needs of policy makers and users.   The seminar generally agreed on the
importance of this and that at the present the available human resources are
inadequate for this task.

As an example of the types of substantive knowledge that are important for
officials of statistical agencies to have, Julius Shiskin, Commissioner of the
Bureau of Labor Statistics (BLS) of the U.S., explained that BLS not only
needs someone who is very familiar with how price statistics are computed, but
also they should know what the typical behavior of prices is.   For employment
statistics, it is important to be familiar with the behavior of unemployment
and the patterns of unemployment associated with different economic conditions.   This substantive knowledge or subject area analysis results in
harmonious relations between the statistical agencies and the policy-makers.
Also, by following through on the acquisition of this kind of knowledge, by
following through on the user needs, and by following through on this type of
analysis, statisticians can maximize the use of the data.

The U.K. Delegate added that in the previous year they had had problems with
their trade statistics which could have been avoided had the statisticians had
more substantive knowledge.   They have now established a rule saying "any
figure that looks interesting is almost certainly wrong."   They encourage
their statisticians to look for interesting figures.

Will the System Work?

Mr. Fellegi of Statistics Canada explained that in developing this approach
they did not have in mind an enormous single data file containing all statistical information nor an all-emcompassing data bank.   If this were the goal,
then the task would be "fantastic".   What they had in mind was more modest;

what they were suggesting was a system for establishing procedures, policies, and standards which would facilitate access to and use of data that is already available within the statistical organization. The alternative--not having a systematic approach to statistical operations--is not very desirable and not very efficient. For example, the ad hoc approach to electronic data processing is vastly more expensive than a systematic approach. Their view of the ideal statistical system is not one that will be achieved overnight, but an incremental approach built up on a day-by-day basis.

Is there a danger of the system controlling man? Basically, the level of development and the direction of development of a statistical system is determined by the level and social development of the country. There is no real danger as long as the controls, as described in the Canadian paper, are functioning. The function of the control acitivity of the statistical system is to harmonize the statistical system with its environment.

The discussion of the systems analysis approach as applied to statistical systems pinpointed the fundamental question of the seminar: what will be the future dialogue between the statistician and his environment? It is difficult to define this relationahip; but it is essential to define it if the statistician is to have an influence on the environment.

Computer Hardware and Software: It s Use in a Central Bureau of Statistics

J. Kazimour

President, Federal Statistical Office, Czechoslovakia

INTRODUCTION

In the three decades which have passed since the end of World War II, the social and economic development is characterized by an extent and importance which cannot be compared with any other period in the history of humanity. These changes have, in particular, the following remarkable features: an abrupt development of science and technology which has also affected consumption, an accelerated increase in the number of inhabitants, the generation of higher social systems, the growth of international integration.

This was immediately reflected in the increasing part played by and the demands for capacity and quality of statistical information.*

The consequences this development had for statistics are the following:

The extent of the data, which statistics are required to provide, has increased

The call for quick decisions has increased and this is reflected in the demands for rapid supply of data on changes

Methods of statistical analysis and econometrics have progressed and this calls for further data, hitherto unrecorded, and more computations

Requirement for flexibility and quick reaction to unforeseen events in management, as well as the high degree of complexity of objects of management, call for the possibility to process unplanned and single-purpose analyses for a concrete customer (a government institution, a scientific or research establishment, etc.)

Meeting the above requirement and also the requirement of providing information quickly in turn demands that the subject analyses can be carried out using data, obtained systematically in the past. This makes it necessary to preserve the data in a form which would enable it to be classified at any time according to the aspects of the analysis being required

---

*The Organization of National Statistical Services: A Review of Major Issues, Mat. of Statistical Commission, nineteenth session (E)CN.3(495), para.5.

D

In meeting all the conditions, mentioned above, it is necessary to avoid overloading the intelligence units and, on the contrary, to try and diminish their load

Efficiency and reliability of the statistical service is required

Security of statistical data is required.

These problems can be solved by computer equipment, at present represented by computers of the third generation.

It may be assumed that, beginning with the year 1980, its development will tend towards automated macro-systems, an important part of which will be formed by automated data systems, including statistical information systems.

Thus, the trend will reflect a gradual transition from the use of individual computers and specialized computer systems to extensive national and gradually also international computer networks. This process will result in the growth of the complexity of the problems of the whole information system and its control.

FORMATION OF AN AUTOMATED INFORMATION SYSTEM UNDER CONDITIONS OF STATE STATISTICS

The development of data systems should be purposefully controlled and co-ord - nated with the level and requirements of controlling the national economy with due respect to its other functions. As optimum one can consider a state in which information, designed for governing authorities, corresponds to their requirements depending on the degree of their authority and responsibility and also when the statistical information system corresponds to the contemporary model of management. The design of this system will systematically exploit the principles of the theory of information, the theory of controlling systems and related scientific disciplines. Drawing on its analogy with live systems, a data system (statistical or general) represents a nerve-system of control, in which the statistical system, concerned with the summary of all data on the state and development of the society, fulfills an insubstitutable role of the highest priority.

Effective control of the economy is more demanding as regards the information system, particularly with a view to the transformation of economic structures into organizational structures, a unified meaning of balance-of-payment indices, linking and compatibility of classification in economic and technical documentation, etc. Meeting all these requirements calls for the establishing of a complex kit-type classification system which must not only serve the purposes of state statistics, but must also be comprehensively applicable to economic planning, to the tax system, in areas of home and foreign trade, at factory level, and must also create conditions for the mechanization and automation of administrative operations in the central organs and institutions.

Experience in the past has shown that the agenda approach to data processing does not exploit all the possibilities which modern computer equipment provides. The solution to this problem is the replacing of individual data processing by the establishing of data sets and the common exploiting of data bases. This approach emphasizes the necessity of systemwise treatment of data

processing and exploitation, i.e. the establishing of a statistical information system, as well as the forming of a new system of codes founded on
strictly logical and multilateral classifications and on the innovation of
indices.

In innovating the indices the following principles have to be adhered to:

(a) In the area of rendering the information more analytical: concentration
on improving the informative capability of the data on social and economic
phenomena and processes so that they provide an objective background for the
decision process. This calls for the data sets to be treated for factories,
trusts and concerns on the basis of certain criteria, expressing, e.g., the
fulfillment of the plan, the dynamics and effectiveness of development.

(b) In the area of exploiting statistical and mathematical methods (this
concerns in particular correlation, regression and factorial analyses which
proved most effective in investigating larger numbers of factors): improvement of the methods of time series analysis in order to improve research into
the social process and provide tools for short- and long-range forecasts of
the social and economic development, and the use of mathematical economic
models for investigating the process of reproduction as a whole and exploiting
structural analysis to improve control of the balance of the national economy.

(c) In the area of differentiation of statistical information (this concerns
in particular the manner, form and contents of the data being provided with
the purpose of making available data already evaluated): provision of a more
profound knowledge of the structure of the phenomenon or process being evaluated from the organizational, territorial, temporal (operative data and information of a long-range nature) point of view and as regards their content
(authorities on a lower level require broader and more detailed information,
authorities on a higher level - above the ministerial level - require comprehensive data summaries for their work). A wider application of progressive
methods in statistics is more demanding as regards the extent of computations
which can only be made by using corresponding means of computer technique.

The state statistical information system reflects the social and economic
reality of the state. The phenomena and processes in this reality are always
directly or indirectly linked and they form an organic whole. The information
system as a representation of this reality should reflect the essential properties of the system being represented, a knowledge of which is required for
optimum control and harmonic development of the society. Therefore, all elements of information and integrational relations must be reflected in the project of the information system.

Of special significance for the integration of the statistical system is the
establishing of an information language which is closely associated with
utilizing classifications, registers and nomenclatures.

"The integration of information systems" should be understood as a process,
the purpose of which is to establish the closest relation between two or
several information systems, which will enable them to be mutually interconnected in order to provide the appropriate data according to the requirements
of the querying information system. Integration, thus understood, will enable
the data of one information system to be utilized by other information
systems, without having to duplicate the determination, collection and storage
of these data. It allows the principle of the uniqueness of occurrence of a

datum to be adhered to. In this sense the significance of so-called administrative registers, containing individual data on which the statistical information systems will be able to draw in ever increasing volumes, will be
enhanced.

The catalogue of data, the information metasystem, common to all participating
information systems, which determines in which information system the required
datum is to be found and in what way it can be retrieved, is important for integrational work.

THE COMPUTER AS A TOOL FOR CREATING, IMPLEMENTING AND REALIZING THE SYSTEM APPROACH UNDER CONDITIONS OF STATE STATISTICS

The automated statistical information system, as opposed to the agenda automated treatment, is more demanding of the technical as well as programming
aspect. Of the various characteristics of the automated statistical information system it is possible to mention the following:

(a) retention of information consisting of a large number of hetero-
geneous statistical units, described by a large number of characteristic
data in various divisions and combinations (organizational, spatial and
temporal division)

(b) constant periodic increase of the data fund and difficult maintenance of the information base up-to-date during organizational changes

(c) existence of a large number of recordable relations among individual
data

(d) necessity of frequent manipulation with extensive data sets (retrieving and classification)

(e) retention of ordinary and secret information in one data base and
necessity to implement effective security measures with regard to the
appropriate parts of the data fund.

(f) necessity to work with several procedural languages of a higher
level and with several operational systems.

(g) necessity to use a non-procedural (consumer) language enabling the
queries of consumers to be simplified as much as possible.

The nature and volume of the statistical data, as well as the specific problems associated with processing them, are indicative of the specific requirements as to the programming equipment of the computer system. The statistical
data for purposes of complex analysis are usually located in different sets.
These sets are frequently located in different territorial localities, have a
different format, etc. Practical statistics call for the collection and processing of relevant statistical data from statistical sets organized and dis-
tributed in this manner. From this point of view the requirements integrated data processing in statistics is much more important than other
fields of data processing.

In view of the large volume of statistical data, considerable attention must also be devoted to efficient storage of data, to their accessibility, topi-. cality, and last but not least also the quality of the storage of the statistical data.

The programming equipment of a computer system for statistical purposes should provide:

In the data collection stage--

(a) singular automatic control and corrections of input statistical data,

(b) necessary aggregation of these statistical data which will not be required for further processing in a de-aggregated form.

In the data processing state--

(a) simple data organization (in the data base) satisfactory for periodic processing of statistical data, as well as for unforeseen requests for statistical information.

(b) expression of the relations between statistical data in different fields of statistics (finance, industry, manpower)

(c) using suitable techniques (data compression), reduction of the volume of stored statistical data

(d) meta-information tools for stored statistical data (systems of registers, catalogues and vocabularies, etc.)

In the data presentation selection stage--

(a) suitable language tools for the programming and non-programming consumer (statistician, analyst, etc.)

(b) simple tools, tabulation of statistical data and table formats with the possibility of storing and retrieving standardly produced tables

(c) reduction of the volume of the produced statistical outputs by a suitable presentation of the fundamental statistical data (e.g., graphical output)

As regards the anticipated development of the programming equipment, the following can be mentioned:

(a) development of specialized systems of data base control for use in statistics

(b) development of territorially distributed (distributive) data bases with a possibility of storing and retrieving any of them

(c) development of tools (compiler generators, automatic testing and documentation programmes) as instruments of generation of programming systems

(d) development of new means of communication designed for a wider circle of consumers--non-programmers (problem-oriented languages, interactive and dialogue languages)

(e) development of meta-information tools (vocabularies and data directories), which will serve the analyst to determine the information content of the
data fund.

What progress of computer equipment can be expected within the next ten years?

Within the next five years, elements with transmission rates of the order of
10-8 secs will probably be available for computer logic and memory circuits,
i.e. 10 and 50 times faster than those used now. In commercial computers
parallel circuits will be used to an increasing extent with the purpose of
cutting down the delays which now occurs in interconnected data transmission
and manipulation. This parallel interconnection will be realized by means of
special processors, designed for controlling inputs and outputs, data sets,
memories, interruptions and similar functions.

A set of micro-instructions in these processors, forming the computers, will
facilitate the fundamental division of the functions and secure the system
functions which are now being carried out by the operation system, e.g., it
will enable several regimes to be run for the emulation of previous programmes
(with the corresponding operational system) as well as to cater for more consumers at the same time, requiring several processing regimes; it will secure
the separation of the input and output functions, data transmission and control of sets from the computers themselves; automatic memory and data-set
hiearchy control at a symbolic level for faciliting programming, easy use and
securing data against misuse (by means of appropriate tables, connected with
individual processing regimes), etc. The computer software will cater for
some of the other functions, connected at present with the activity of the
operational systems.

Therefore, the computers of the next ten years will consist of processors
which will automatically carry out the micro-programme functions controlled by
the operational system. The run of the computer will be determined by their
built-in processors. Three levels of built-in processors are defined, the
processors of the third level corresponding to the central computers of the
present intermediate and higher performances, but will have the same perfor -
ance as the present largest computers. They will be used individually in
computers with one processor, designed for batch processing, like to-day, and
in multiple systems with multiprocessing operational systems and in large
systems with a high automatic resistance to failure. The manufacture of high-
performance computers, higher than that of third generation computers, is
anticipated. These computers, however, will only be manufactured for large
scientific systems and will not be available for commercial systems.

Practically, one may anticipate the use of micro-computers with processors of
the first level in intelligence terminals or satellites; by 1985 one may
expect the use of micro-computers for independent systems. The mini-computers
will be capable of processing data. A central unit will consist of one processor of the second and third level and as many as 20 processors of the first
level which will cater for the operation of the peripheral equipment. The
main purpose of these computers will be use in interactive applications requiring large memories (a capacity of 0.2 to 0.5 MB of the main memory and
4 MB of the auxiliary memory is anticipated). The monocomputers will mainly
be designed for consumers who will require processing of large volumes of data
in batches, although the possibility of interactive processing will be available. In these systems it is expected that one central unit of the third

level, two to three processors of the second level for controlling high-capacity units and an adequate number of first level processors for slow peripheral units, will be used. The capacity of the main memory is anticipated at 2-4 MB, the auxiliary memory at 30 MB. Multiprocessing will not be used: however, from the point of view of performance a fixed division of the memories and an interactive virtual regime are anticipated.

As regards external memories, gradually introduction of new memory system is anticipated together with the improvement of the characteristics of the current magnetic memories. For example, the possibility of increasing the recording density by a factor of 40 is being assumed, which will mean much lower costs per unit memory bit by 1983. With a view to the difficulties with the access time of magnetic memories it is probable that in 1985 the same hiearchies of auxiliary memories will still be used.

In the forecasts concerning input and output devices for work in batches a whole spectrum of these devices with various speeds and functions is mentioned, as well as the conjugation of the individual equipment into stations of which networks can be assembled. With the two fundamental types the line printout is being combined either with punch cards or a magnetic external memory. Two optional auxiliary devices are being forecast: optical pickups and microfilm output printers. The introduction of effective, slow and inexpensive OCR stations will be responsible for the development of complicated kit-built devices with many input media between 1977 and 1985; these will combine the slow operations in batches, picking up the characters and terminals for data input from the keyboard.

The small computers and terminals will usually be equipped at least partly for data transfer, because most of them will be connected, at least for part of the time, to communication lines, and the costs of adding another first level processor for this purpose are insignificant. In connexion with the transfer of data, packet switching is being considered to a broader extent, to replace leased or commuted lines. The technique of packet switching was developed especially for improving transfer data services and enabled power, corresponding to the requirements of connecting the terminal with the computer and of connecting computers to one another, to be fed into the networks. The characteristics of these networks are in particular: fast response, high reliability, very small error rate, dynamic distribution of the transfer capacity, proportionality of costs to the volume of transferred data, and improved transfer capabilities. However, the exploitation of the advantages of packet switching depend on the construction of extensive transfer networks, which require high investment costs.

In the area of software one may expect development in operational systems, the functions, of which, to a large extent, will be taken over by micro-programmes. The principal functions, which will be preserved, e.g. planning jobs, distribution of unsharable units, error reporting and renewal after failure, will be catered for by relatively simple monitors.

By 1985 computer systems will control automatically the activities of external devices including libraries on tapes and discs, planning of external jobs, recording and invoicing of computer time. They will also automatically protect data sets: the system of data set control will govern symbolically the access instructions and the control system (inaccessible to the consumer) will record every contact with the data set.

The pattern of the data bank control systems calls for a supervisor of uniform data control, a question-answer module, a data definition processor, a
reorganization backup module and a statistical performance analyser. The
future data control supervisors will devote more attention to securing the
correct function of the data bank. This will include the possibility to code
certain relations into the data bank definition and control of agreement which
must exist between the individual records or fields. At each change of record
or field these agreement checks are called and if the prescribed conditions
are not satisfied, the operation is not carried out. The function of the
question-answer module will include general routines for contact with the
supervisor of the uniform data control, e.g., the possibility of specifying
various criteria for record seeking, to carry out fundamental computations
with the approximate data and provide specialized or singular data bank dumps.
Further, assuming access authority has been obtained, the consumer will be
capable of writing simple programmes which will change the data bank; this can
be utilized for data input operations or simple transactions.

The data definition processor is used to define the structure of the data bank
and other characteristics such as access limitation, etc. A higher degree of
data independence, which concerns the separation of data definition and their
processing by application programmes, is becoming more widespread in the
development of data control systems. The programmer will require less knowledge of the data bank structure and will have more freedom in reorganizing or
changing the data bank structure without affecting the work of the individual
programmes. This will help in securing independence of physical media, the
possibility of adding new types of records to the data bank, adding new fields
to the existing records and changing the relation between the elements and
records. Further flexibility exists in the arranging of the data structure
and methods of access.

All the development trends of the hardware and software of computers, mentioned above, indicate that the transition from agenda treatment of statistical data to the gradual establishment and introduction of automated statistical information systems is justified.

THE CONSEQUENCES OF ESTABLISHING AND IMPLEMENTING AN AUTOMATED STATISTICAL INFORMATION SYSTEM IN STATE STATISTICS ACTIVITY

In introducing any automated system, demands are being put not only on changes
of the technical side of this system, but also on changes of the organizational and professional structure of the existing system. Therefore, an automated
statistical information system, established in the object of state statistics,
will affect its structure. If this system is to provide compatible information, a uniform method of producing statistical indices must be unexceptional-
ly valid within the whole state statistics. This does not hinder the development of these indices in the content of the statistical system, but supports
integration within the system and its neighbourhood. However, this centralization does not mean that decentralized processing is to be prevented: on
the contrary, if the preparation of the data for processing is to be as fast
as possible, the checks of erroneous data corrections must be made operative-
ly, and the processing must then be as close as possible to the level of
organization at which they can be best executed and exploited.

The nature of working activities will necessarily change under the conditions of operation of an automated information system, and this will be reflected in requirements regarding the qualification structure of the workers. This system will cause a gradual shift of administrative check work of a stereotyped and less demanding nature to the computer.

Some of the existing professional groups will have to be modified to enable them to function effectively within the framework of the automated information system. At the same time new professions will be created, e.g., custodian of the data base who will be responsible for the maintenance and integrity of the data base.

The automated statistical system will also affect the level of qualification of statistics analysts. They will be required to increase their knowledge in the area of data processing, particularly as regards the possibility of utilizing standard programmes from the computer library, as well as other output methods using terminals, displays, x-y recorders and other equipment. It is here where the differentiation must be sensitive in determining the most suitable qualification which cannot remain just theoretical, but must represent real functional education.

Attention must also be devoted to workers at the managerial and conceptual levels, because these must affect the forms of securing the individual phases of the work with data.

Similarly, appropriate attention will also have to be devoted to the education of the final users of the statistical data in order to establish efficient communications between them and the automatic statistical system.

The establishing of an automated statistical information system in the form of a computer network, equipped with a bank or banks, and capable of co-operating in automated operation with other automated systems and with a wider circle of consumers, equipped with computers or terminals (prospectively by 1985), will require not only considerable labour and financial expenses, but also a long time for preparation and execution by stages. The actual realization, estimated around 1985, requires intense preparation already now.

The realization will call for the following fundamental problems to be solved:

(a) the large size and the necessary exactitude of the project, requiring the co-operation of a large number of specialists of several scientific disciplines with the necessary practice in designing automated systems, which is usually outside the possibilities of a single organization

(b) the necessity of stagewise solution of the problem as a whole and sufficient intervals between determining the project plan of the individual stages and their execution. The stagewise solution is a necessary condition because some of the important initial data for the subsequent stage can only be obtained by practically realizing the previous stage

(c) the availability of considerable financial means, usually also outside the scope of a single organization

(d) the provable economic effect of realizing the project.

The exactitude of the project of an automated statistical information system from the point of view of manpower and costs, as well as the difficulties associated with securing the effective operation of this system, realized in an independent, single-purpose computer network, will probably call for a different form of execution of the system in most cases. It seems that the most suitable is the designing of nationwide computer networks, already being implemented, capable of assuming and economically executing the functions of several information and management systems, including extensive services for a large number of users of the system. This network has considerably lower demands as regards design work of newly included systems. The task of the user will then only be to organize his data base according to the general principles valid for the given network and to work out a system of those pro-grammes which are not available in any of the network computers. This of course does not preclude the possibility of the central bureau of statistics owning its own computer or computer sub-network, which will become part of the nationwide computer network, will make use of the services of the network for its own purposes and, on the other hand, provide services (data) to its other users through the network.

The organizational structure of the central bureau of statistics will mainly be influenced by the selected alternative of processing (centralized or decen-tralized), use of a centralized or decentralized data bank and, possibly, the function of state statistics as an automated statistical information system. This will change the data flow between the national and regional centres and the work of these centres. However, in all cases the transition from repeated (stereotype) operations to more complicated (analytic and prognostic) opera-tions will continue, which will be especially reflected in the stage in which the conversions from written documents to data carriers, used in automated processing, will be carried out in a more effective way (a more massive implementation of reading devices), or in which they will become the by-product of the usual function of the investigated objects.

However, these increased demands, on the other hand, will have numerous ad-vantages for the workers of the central statistical bureau and other consum-ers, as regards easier and faster access to the data, as well as their infor-mation content and more lucid arrangement of the output data (graphical outputs by means of x-r recorders). The co-operation with automated statisti-cal information systems in the form of a conversation regime will be just as important.

Problems of protection against destruction of the data in the data carriers will be particularly important, be it due to unintentional erasure during their use, or due to the ageing of the magnetic record. This will require special measures in archiving the data carriers (duplicate copies, their periodic renewal and other measures). Of the same importance are the problems of secret data, data of particular state importance and questions of the judicial validity of the data provided by the automated statistical informa-tion system.

CONCLUSION

The introduction of computers into state statistical services causes a con-siderable change in the organization of practical statistics. Statistics are making a large jump from presenting source data (documentation function) via

agenda and single-purpose processing to the formation of a statistical
information system, enabling a comprehensive statistical information system
to be established, using data bases including time-series, methods of mathematical statistics and other tools of computer technology, its main advantage
being the determination of the causal relation of the development of society
and the national economy, to analyse the factors of this development and
elaborated prognoses of further development.

At the same time, one should bear in mind that computers and other devices
for collecting, transferring, presenting and processing data are only an aid
to man, are his tool.

Not even in the state statistical service should the use of computers be an
end in itself. In this connexion it would seem suitable to remind the reader
of the words of Academicain Glushkov, Member of the Academy of Sciences of the
USSR, who said, with regard to the problems of introducing automated systems,
that the advantage of automated systems is primarily in that the quality of
work of the controlled object will improve by one or more degrees. The
fundamental effect of automated systems is not to release people from the
system of management, but to bring about an abrupt improvement of the quality
of management.

The international co-operation of specialists has also contributed to the
solution of the problems of using computers in statistics. The mutual
co-operation of central statistical bureaux should also be developed in future
in superior forms. It is assumed that the exchange of experience will continue to be organized as part of the work of the working group of the Conference of European Statisticians for Data Processing and at regular ISIS seminars. The co-operative research programme for establishing and developing
computerized information systems, which should receive support from the statistical bureaux of the ECE countries, as well as the organs of the United
Nations, is a new and prospective form of co-operation conforming to the
spirit of the provisions of the Final Act of the Conference on Security and
Co-operation in Europe.

Discussion

The discussion of the Czechoslavakian paper on "Computer Hardware and Software: Its Use in a Central Bureau of Statistics" indicated a general agreement among the delegates on the role of computers in society. They agreed
that the role of computers will increase. Central statistical offices have no
alternative than to accept the role of computers since computers provide such
a powerful tool for creating new statistical systems. Central statistical
offices must shape their roles and functions to take advantage of computers.

Role of Chief Statisticians

The challenges created by the increasing role of computers are quite substan-
tail. These challenges must be met and borne by the chief statisticians. It
is they who have to come to grips with the fact that electronic data processing is going to have a fundamental impact on all statistical organizations including the staff of the central statistical office and on the work enviroment
of all the people. If chief statisticians do not concern themselves with the
relative needs of the organization, then they are going to feel the impact in
terms of a very troubled scene. It has to be a policy decision at the very
highest level concerning the direction electronic data-processing is going to
take in the statistical agency.

The chief statistical officer, in the future, is going to have to define the
goals and objectives of the functions of the central statistical office more
clearly, and then explore the methods for achieving them. In the past, statistical programs were not planned from the perspective of implementing large
scale statistical systems; they devel .u in a more traditional setting of an
ad hoc approach. Now, statistical officers must clearly define their priorities and then define their goals and objectives in the priority areas.

It is the chief statistical officer who will have to deal with the problems in
upgrading large-scale statistical systems. This is difficult for several
reasons. First, this requires substantial resources; the on-going work load
of a statistical office is so heavy that the available human resources are
quite limited. Secondly, the budget process only looks two or three years
ahead, whereas the planning process for a program, such as a decennial census,
requires about 15 years of planning. Thirdly, it is extremely difficult to
match the new technology and the planning process with the ongoing work program while avoiding total collapse between the realization of the idea and the
availability of the new data.

Software Development

The chief statistical officer is also going to have to find better ways to use
the hardware which will be available; the constraints are not in the hardware,

but in the software which includes programming, organization, and people. Today we have practically reached the speed of lightning in the design of hardware, but the design of the software will limit the use of the hardware for
the next twenty to thirty years. Unless chief statisticians push the development of software now, they will not be able to use the hardware that will be
available in the future.

Probably the most important aspect of software is the human aspect. In considering human resources in relation to the computer, the problem of greatest
concern was the integration of the computer specialist with the statisticians.
The problem is to integrate the EDP personnel with their different training in
the personnel as a whole, so that they can fully collaborate with the statisticians to develop and improve the system. The computer specialist does not
have statistical training; the statistician does not have computer training;
but he may have ideas about electronic data processing which are different
from those working in the field. There will be conflicts. Also, it is the
nature of the computer specialists not to deal with uncertainty, whereas the
statistician has learned to deal with uncertainty. So there is a conflict in
that the computer programmers are very deterministic and the statisticians are
not. Here again, it is the chief statistician who should deal with and have a
concern for people and their careers.

Data dictionaries, another aspect of software, are a way of protecting the investment in the data already collected. They allow the data to be used on a
longer term basis and bring a standardization to data that is otherwise not
possible. Without such data dictionaries there would be a long lead time and
high cost in bringing about changes.

One participant of the seminar suggested that the best way to get more efficiency and better performance out of the hardware is to allow things to take
place in parallel. He suggested that statisticians might look at the fundamental flow chart of a statistical system presented in the Canadian paper and
ask which of these functions can go on at the same time. What can go on
simultaneously in the specification of the problem?

International Cooperation in the Use of EDP

The Seminar also expressed the hope that its participants could organize a
greater amount of coordination among the statisticians of various countries in
using electronic data processing. The U.S.S.R. delegate explained that within
the communist countries they have a permanent working group which coordinates
statistical bodies within the socialist countries in the application of computers to improve current practices in the use of computers in statistics.
This sort of exchange should continue in other forums. He suggested a Conference of European Statisticians seminar similar to the present one in which the
use of electronic data processing is discussed in-depth. In this connection,
the Director of the ECE Statistical Division suggested that as the work program of the Working Party on Electronic Data Processing for the next year had
only been decided in part, the Conference at its next plenary session in June
might wish to consider this as a topic for the Working Party's program. He
added that the seminars organized by the Computing Research Centre in Bratislava also include a cooperative research program on the use of computers for
statistical purposes.

Priority Setting in the Coming Decade (Survey Linkage and Integration)

J. W. Duncan

Deputy Associate Director for Statistical Policy, Office of Management and Budget, United States

Introduction

The preceding papers in this seminar have considered the environment for government statistics in the coming decade and some concepts for dealing with that environment. The discussion has focused mainly on the challenges which are likely to be associated with changing views of the public concerning appropriate functions for government, the utilization of improved computer capabilities and other technological changes, and the growing difficulties with respondents who supply the basic data for government statistics. At the same time, demands will increase for more and better statistics to meet the needs of business decisionmakers, academic and research interests, and governmental policymakers.

Many of these developments have already emerged in the last ten years. The main response of Central Statistical Offices has been to develop a proliferation of special-purpose studies and surveys to meet these individual demands for specific information and statistics. Many governmental statistical agencies are faced with limited budgets, a need to reconcile conflicting data series, and consequently, strong pressures to define priorities among the widely diverse statistical programmes which presently exist and which are contemplated.

This paper first focuses on a series of propositions which describe the present state of affairs facing most Central Statistical Offices. It then discusses approaches to setting priorities. In view of the pressures on statistical offices that are discussed in the first part of this paper, it is unlikely that the entire range of national statistical programmes can be fully rationalized and that an integrated set of priorities can be applied to all statistical efforts. Distortions and imbalances are likely to remain. However, this paper outlines an approach for setting priorities in selected categories of general-purpose economic and social statistics.

For economic statistics, the national income accounts provide a useful basis for setting priorities among the statistical series that serve as the basis for estimates of the various components. By analysis of the differences between preliminary estimates and final estimates based upon more complete data, it is possible to identify those areas which need improvement in relation to current estimates. The costs of those improvements can then be compared on the basis of resulting reduction in estimating error to establish priorities concerning individual contributing series.

In the area of social statistics, an integrating framework does not exist.
Hence, the task is much more difficult. As the subtitle implies, this paper
proposes an approach for integrating major social surveys. The approach is
possible in the United States because a recently authorized mid-decade census
offers a flexible potential for comprehensive integration of special-purpose
social surveys for statistical purposes. In effect, a mid-decade "statistical
effort" is proposed in which a series of "nested" surveys would provide a common set of demographic characteristics to serve as the basis for statistical
matching of detailed information collected concerning specific characteristics. With this approach, priorities for special-purpose social surveys would
be established on the basis of their contribution to the overall integrated
demographic and social statistics system.

It is intended to set forth a strategy which can be helpful to Central Statistical Offices in meeting the problems which have been addressed in earlier
sessions. While this paper offers a potential approach to the difficult task
of setting priorities for major governmental statistical programmes, it is
not intended to predict the processes which will be implemented in all countries nor does it attempt to foresee the new forces that will impact on statistical programmes in the decade ahead.

The Difficulties of Priority Setting

For the past several decades, the Central Statistical Offices of most countries have attempted to outline rational programmes of statistical development. These efforts have required explicit allocation of scarce financial
and professional resources to the tasks which meet the most important needs.
In a widely distributed report, "Setting Statistical Priorities", the Committee on National Statistics of the National Academy of Sciences in the United
States recently discussed the problems of priority setting.

The essential recommendation of this report was that "the assignment of
priorities among data packages and programmes should involve an explicit consideration of anticipated benefits and costs" The Committee on National
Statistics report discusses approaches for overcoming the limitations of
classical cost/benefit analysis techniques when addressing the issue of setting statistical priorities. The basic recommendation was backed up with
supporting recommendations that encouraged:

   1. The establishment of technical groups to undertake such analysis

   2. The expansion of research on measuring benefits and costs of data
   programmes

   3. The centralization of statistical activity for specific functional
   areas so that there would be an informed determination of needs and
   priorities.

The supporting recommendations or related actions are required because the
Central Statistical Offices are faced with the practical difficulty that the
benefits of individual statistical activities are frequently hard to identify,
especially in terms of secondary and tertiary impacts. For example, the use
of national income accounting estimates is so widespread among governmental
and non-governmental groups that it is virtually impossible to measure the
benefit of improved estimates of national income or to evaluate the detri-

mental effect of decisions based upon inadequate series. This task is made
more difficult since the relative importance of different sectors of the
economic structure varies considerably depending upon the stage and the
particular characteristics of current business cycle developments. Likewise
the Central Statistical Office is often unable to measure the full cost of
gathering various statistics since the direct costs to governmental agencies
do not reflect the burden of the recordkeeping requirements or reporting
demands on respondents who supply the data. If the required information is
not readily available, these additional costs can be very high.

Other techniques used by the Central Statistical Offices include analysis of
programme management impact*. Under this approach the statistical programme
is considered as a part of the programme management associated with specific
government programmes. Another approach is the recognition that certain
governmental priorities determined by legislation demand priority statistical
activities. These legislated priorities often absorb a significant portion of
the statistical budget without having any relationship to related or allied
statistical programmes.

Most of the development of the statistical activities has been a product of
incremental improvements to individual series after they have been established
or by episodic emphasis on statistics in areas of immediate or growing public
concern. Before proceeding to discuss means of improving this process in the
coming decade, it may be helpful to set forth some propositions concerning the
considerations which face Central Statistical Offices.

PART I: SOME PROPOSITIONS CONCERNING PRESSURES ON CENTRAL STATISTICAL OFFICES

In establishing priorities among statistical programmes, one must recognize
that there are several basic pressures which are continually placed on the
statistical policymaker. Some of these are:

1. Every statistical series has users who exert pressure for continuation of
ongoing efforts

2. Legislation is typically designed to focus on specific (but inconsistent)
target groups

3. As policymakers and programme managers demand more precise data, the
impact on the providers of data multiplies in complexity and magnitude

4. Both crisis management and new policies yield pressure for quick response
by the statistical agencies, reducing opportunities for adequate co-ordination

5. With growing sophistication of social service delivery, there is increasing demand for greater geographic and demographic detail for established data
series.

A brief review of each of these propositions is presented below as background
to the concluding section which outlines one approach to improved priority
setting.

Every statistical series generates users. Once a statistical series is
published by the Government, a variety of policymakers and business and
academic analysts begin to determine what utility individual estimates have
for them. Inevitably, certain insights are drawn and, over time, a continuing
analysis of those series is undertaken by affected groups. Even though the
demographic or other conditions relating to the series may change considerably, there remains considerable pressure to maintain old and familiar series.

For example, the concept of a family composed of a working male with a dependent wife and two relatively young children, as descriptive of the typical
household, may have characterized a large number of households at one point in
time. The emergence of secondary and tertiary wage earners, combined with
greater variation in living arrangements, has resulted in a considerable
change in the character of households**. Nevertheless, the old stereotype
continues as a statistical and social reference point in the United States
being used for statistical series concerning real income levels and as a basis
for estimates of social programmes. A Central Statistical Office which sets
out to define a household that approximates current household relationships is
likely to encounter considerable resistance from users who are more familiar
with the older definitions. Thus, users of existing series resist their
abandonment even if priorities, concepts, or realities have changed.

Legislative focus on specific target groups. Much legislation is developed to
deal with specifically defined problems. For example, in the field of education, one may wish to establish programmes to deal with the special problems
of poor families, to enhance opportunities for gifted children, etc. These
legislated activities frequently generate programme-related statistics or
macrostatistics which are intended to describe the problem being addressed by
the programme and the impact of the programme on improving the condition of the
specific target group. The affected groups may frequently overlap. For
example, a gifted child may be a member of a poor family and thereby qualify
for both of the programmes just mentioned. Over time mandated statistical requirements associated with measuring each programme independently lead to
overlap, duplication, and confusion. The statistician's flexibility for (a)
reducing the volume of required information in setting priorities concerning
detail of coverage and (b) using general-purpose definitions is restricted
by such legislation.

----

*These additional categories are discussed more completely in a paper presented
at the 40th Session (1975) of the International Statistical Institute in
Warsaw, Poland. See Joseph W. Duncan, "Developing Plans and Setting Priorities
in Statistical Systems," ISI Contributed Papers, pages 245-253.

**In the United States, "for example, in March 1975, only 21 per cent of the
71.1 million households in the United States consisted of husband-wife families in which the husband was the sole breadwinner." Source: Bureau of the
Census, Current Population Reports, Series P-20, Number 291, Table B (for
number of households) and Bureau of Labor Statistics, Special Labor Force Report Number 183, Table M (for number of husband-wife families with the (male)
head employed and the sole breadwinner).

Reporting requirements grow in complexity. As individual demands for information are generated, they may appear to be perfectly reasonable individual
requests. However, the cumulative effect of programmes with different reference dates, with different industry coverage, or with different definitions create a serious impact on the respondent. His reluctance to meet the
unique requirements of individual statistical programmes increases with each
request. Hence, what may appear to be appropriate from a limited programme
management perspective, may be highly inappropriate from the standpoint of
developing compatible statistical results that minimize burden on the respondent.

In the United States this problem is further complicated by a recent tendency
to substitute statistical collection for direct regulation. The philosophy
is that if institutions, individuals, or corporations are required to report
data concerning such activities as equal employment hiring, the requirement
to submit data will force institutions to reconsider discriminatory practices,
having an effect similar to direct regulation. The reporting burden associated with such universal inquiries generates frustration and resistance among
the providers of statistical information and may, in fact, be detrimental to
the accomplishment of the original policy objectives. In summary, as data
requests expand, the effect on respondents multiplies in impact, thus generating widespread resistance.

Crisis management and new policy generates pressure for quick response; little
coordination. A recent example of this problem in the United States related
to the energy crisis of 1974, in which there was an immediate demand for
improved information concerning sources and uses of alternative forms of
energy. The governmental role in energy regulation was relatively limited
before the energy crisis. Hence, existing data in the United States (and also
in most developed countries) did not meet the demands associated with fully
understanding energy problems and alternatives. With the 1974 energy crisis,
many crash efforts were undertaken to generate new series of data to meet the
immediate policy needs. Unfortunately, this resulted in the establishment of
duplicative reporting requirements and the introduction of conflicting definitions. Statistical programmes frequently originate under such conditions
and, as a consequence, time for appropriate co-ordination or priority setting
is not allowed. In the end, the programmes developed may be less appropriate
than would have been achieved under better co-ordinated conditions.

Increasing geographic detail is needed. As the complexity of social service
programmes has increased in recent years, it has been recognized that there
are substantially different needs for families living in rural areas, small
towns, or urban concentrations. These differences are especially important
as there is a national goal of equity in benefits accompanied by a desire to
shift programme administration to the local areas. Cost-of-living estimates
that apply to a nation as a whole cannot be expected to reflect the unique
conditions of smaller geographic areas. Also, as programme responsibilities
are decentralized to the local level, there is a growing demand for increasing geographic detail in statistics. Since the development of local area
estimates is considerably more expensive than the development of national
estimates, a high percentage of the national statistical budget may be required to develop only a few local level measures.

These propositions concerning the recent context of priority setting are
likely to continue to be important in the coming decade. Consequently, pressures for maintaining special-purpose programmes are inevitable, and it will
be difficult to fully overcome the fragmentation of statistical efforts which
has characterized the last two decades.

PART II: APPROACHES TO THE SETTING OF PRIORITIES

In the following sections separate views will be presented concerning
approaches to the setting of priorities among programmes affecting macroecono-
mic statistics and programmes designed to yield social statistics.

Integration of Economic Statistics

In the 1970's most developed countries have produced an impressive variety of
economic statistics to describe both current and longer term developments with
sufficient detail so that the economic policy analyst can generally understand
current conditions in the national economy. Nevertheless, there are difficulties since comprehensive and accurate measures often are available only
after considerable time delay. Thus, there are frequent revisions in preliminary estimates of national income and related measures as more complete information becomes available.

This difficulty is further complicated by the fact that current estimates
frequently are based on partial information that may be collected using widely
diverse methodological concepts or definitions of industry sectors or consumer
groups.

A vital factor in improving economic information is the standardization of
classifications and definitions for key series. When the input of wages or
materials for one industry sector is collected using a concept different from
that used for estimates to measure output, there are great difficulties in developing accurate estimates of such items as productivity and value added.

The national income accounting concepts provide a refined framework for economic data collection. The key to improved economic information is the integration of data collection efforts that provide input to the national income
accounts with the accounts themselves. Considerable progress has been made in
recent years, but it is anticipated that during the coming decade even further
improvement will be achieved.

In the United States, for example, we expect that the industrial directory
concept (which provides a complete listing of firms by industrial category,
size and location) will lead to the development of integrated samples yielding
consistent estimates of the wide variety of inputs and outputs which characterize economic activity. Further, intensive reviews of recordkeeping
practices in businesses and institutions will lead to improved comparability
in the information provided by respondents. Finally, improved economic models
are already being developed to understand such things as the sources and uses
of energy, the cumulative effect of point estimates of enviromental impacts,
and even such a refined concept as the quality of life, to the extent that it
can be measured.

Setting priorities. In this context, the setting of priorities is greatly
simplified. The national income accounts (NIA) provide a framework for data
collection, and the industrial directory, improved models, and better classifications provide tools for achieving improved economic statistics. Using the
NIA framework, it is possible to identify gaps in current information and to
devise programmes to meet those gaps. Then, by examining the impact of specific NIA item estimates on overall results, it is possible to determine overall benefits to be achieved by the new programmes in relation to the overall
costs of the statistical effort*.

The approach is straightforward. First, an analysis is made of the differences which are observed between preliminary estimates and final estimates of
components of national income and product accounts. The sources of the preliminary estimates are then examined to determine causes of error and to
identify alternative collection procedures. It is possible to set priorities
among suggested improvements by comparing the contribution to error reduction
in overall estimates of national income with the cost and reporting burden
associated with the proposed improvement.

International concerns. In economic accounts the need for greater comparability in trade statistics and measures for cost of living will grow in importance. These needs may not be resolved during the coming decade, however,
it is likely there will be a considerable effort devoted to developing concepts for improving these measures and for standardizing the international
statistics in these important areas**.

Unmet needs. It is unlikely that the national income accounts will be substantially revised in the coming decade although considerable work on this
topic is likely to be undertaken. By the end of the 1980's, however, one
would expect that the basic concepts of the national income accounts would be
revised to more fully reflect the environmental and energy shortage conditions
which are likely to be evident on a global scale. Further, the growing inter-
dependence of multinational corporations and multinational trade compacts will
have created difficult problems for data collection. International statistical organizations, such as the United Nations Statistical Office, must be
greatly strengthened in order to yield significant achievements in these difficult areas and to aid national statistical offices in setting appropriate
priorities.

---

*A recent attempt by the United States to provide such analysis is the Gross
National Product Data Improvement Project. The results of this study will be
published in a few months. (A summary of this project will be distributed at
the meeting.)

**See Kravis, Irving B., et al., A System of International Comparisons of
Gross Product and Purchasing Power, Johns Hopkins Press, 1975.

Integration of Social Statistics

The problems of economic statistics are less difficult than those associated
with social statistics in the coming decade. The national income accounts
have now been under development and in use for over three decades. Hence,
the framework, even though controversial, is widely accepted so that measures
of economic performance are presently being developed in a comparable mode,
e.g., the System of National Accounts (SNA) of the United Nations. Even
though there has been nearly a decade of effort devoted to developing a system of social statistics, there is no comparable theory of conceptual agreement in this area*.

The major area of international agreement relates to the central role of demographic censuses in meeting the fundamental requirements for social statistics. This agreement has recently been augmented, under United Nations
leadership, by a growing focus on multipurpose sample surveys in years between
censuses. Nevertheless, it is clear when one compares the social indicators
publications of various countries that the state-of-the-art of social measurement is relatively primitive and, furthermore, that there is no integrating
concept or philosophy concerning the development of social statistics. Hence,
unlike the economic statistician, who has a major aid to integration in the
national income accounting framework, the social statistician has less overall
guidance.

Nested surveys. In the United States the 94th Congress authorized (in 1976) a
mid-decade census. Although this census will not take place until 1985, considerable work must be done prior to that time. Hence, the concept being
presented here is somewhat speculative since it represents an idea of the
Statistical Policy Division of the Office of Management and Budget, not a
fully accepted point of view of the United States Federal Statistical System.
Further, it should be emphasized that the efforts of other countries may prove
more fruitful and that the results of this conference may generate alternative
views which may be more useful for establishing priorities of Central Statistical Offices for the development of social statistics.

The proposed approach is to utilize the mid-decade census as a "mid-decade
statistical effort" which would develop social and demographic statistics
needed for various formula grant programmes of the Federal Government in al-
locating funds to the different regions, States, and sub-State areas throughout the Nation. This program would refine the special-purpose local area
surveys which were designed to satisfy individual programme needs as discussed
earlier. Most of the existing programmes operate with unique concepts concerning such design features as geographic coverage and definition of covered
households or individuals. Following the 1980 census, it should be possible to
define a set of common characteristics for these individual programmes and to
initiate a general revision of the special-purpose surveys so that they
achieve relatively similar coverage and use similar concepts at the national
level. The mid-decade statistical effort could then be used to develop the
local areas information required for indivdual special-purpose topics.

---

*"Towards a System for Social and Demographic Statistics," ST/ESA/STAT/SER.F/
18, New York, United Nations, 1975.

In effect, this would create a series of "nested" surveys in which a common set
of information would be collected for all populations sampled, but the detailed
characteristics explored would vary greatly among the different survey samples.
This detailed information would be related to the ongoing annual surveys that
presently exist. This would require making sure that the concepts and the
universe covered by the periodic special-purpose surveys are the same as those
used for the mid-dicade statistical effort. It would also require developing
a data file following the 1985 statistical effort in which households with
common characteristics would be statistically matched to yield a statistical
profile of various functional areas even though individual families were not
questioned concerning all specific subject matter aspects. This would provide
analysts with a very rich resource to investigate interactions among various
programmes.

In the United States we have identified a number of programmes which are now
used to give local area estimates and which could be redirected to merge with
a mid-decade statistical effort. For example, the demographic portion of the
Census of Agriculture could be substantially reduced and the mid-decade programme could yield much better data on the total rural population. The metropolitan detail (60 cities covered once every three years) in the Annual Housing
Survey could be dropped in order to obtain housing data for all 272 Standard
Metropolitan Statistical Areas once every five years. When such revisions are
accumulated over the period of the entire decade, our analysis shows that it is
possible to offset the cost of the mid-decade statistical effort. Thus, for
the same total expenditure, a totally integrated set of social and demographic
data can be achieved.

Under this concept, the key to integrating social statistics becomes the use of
common demographic and subject classifications for individual special-purpose
surveys. The tools for developing integrated statistics in the social area include the expansion of multipurpose surveys, the use of the mid-decade "nested"
survey approach, and the development of data files based on statistical matching of data form the mid-decade statistical effort and the annual special-
purpose surveys.

Setting priorities. Under this programme, priorities on the individual statistical programmes will be determined by first relating the topics of special-
purpose surveys to programmatic needs and second by concentrating on expansion
of those programmes which make a contribution to the integrated data set. The
primary impact of this approach on budget setting during the 1980's would be to
restrict the geographic coverage of limited special-purpose surveys and to
utilize the mid-decade statistical effort to generate appropriate mid-decade
estimates for small areas.

Budget resources would be restricted for proposed special-purpose social surveys which did not utilize a sampling frame that is compatible with the mid-
decade statistical effort.

Ultimately the data set based upon statistically matched samples would be compared with decennial census results to determine error terms and needed
improvements, much in the same way proposed earlier for setting priorities on
components of national income.

This approach also places a high priority on research on data linkage based on
statistical matching and the concept of defining linking elements, an effort

which must take place in the early years of the decade. In fact, the 1978
budget for the United States Bureau of Census includes some research funds for
initial exploration of this topic.*

International problems. A current example of this concept in a limited sense
is the World Fertility Survey in which there is an attempt to influence different nations in the use of similar concepts and to refine individual questions to meet local conditions, yet to derive information which is
internationally comparable. The multipurpose survey approach of the United
Nations, at this time, is designed fundamentally to encourage individual
nations to meet many purposes with their sample surveys and thus to generate a
richer base of social data for individual countries.

During the next decade some agreement at the international level must be
reached concerning the concepts and classifications to be used and the overall
framwork within which the multipurpose surveys are developed. Unfortunately,
in the absence of a well accepted "system of social and demographic statistics", it is difficult to implement this approach on an international basis.
Hence, the integrated international approach to social statistics is likely to
be possible only in the 1990's after much more specific experience is gained
with concepts and implementation such as is anticipated in the United States.

Unmet needs. It is evident that the major unmet need in the area of social
statistics is a clearly articulated social theory. Given the differing cultural conditions of individual countries, it is unlikely that a universally
applicable theory can be developed. However, an improved and integrated data
base for individual countries, such as will be possible by the approach outlined above, may assist in the development of concepts that will lead to
improved data comparability.

Summary

This paper has attempted to present a series of propositions which underscore
the importance of special-purpose programmes in the ongoing operations of
statistical offices. At the same time, it has attempted to set forth an
approach for using national income accounting and linked or "nested" demographic surveys for achieving more integrated economic and social statistics.

This paper does not attempt to predict specific developments in the coming
decade, but rather it has attempted to outline an approach which may be in
growing use by Central Statistical Offices in the 1980's in approaching the
difficult task of setting priorities.

In the last analysis, priority setting for statistical surveys must be responsive to the emerging information requirements of our complex and rapidly
changing societies. Many of these requirements cannot be anticipated. We
must therefore recognize, as our first priority, the need to develop a capability for information collection and analysis which has sufficient flexibility
to respond effectively to the information needs which arise in the future.

---

*A paper discussing technical issues to be addressed will be distributed at
the Conference site.

Discussion

Two fundamental questions raised by the delegate from Austria shaped much of
the discussion of the paper on "Priority Setting in the Coming Decade (Survey
Linkage and Integration)"; "How do we decide what to do?" and "What is our
ability to carry through on what we decide to do?"

How Do We Decide What to Do?

The general consensus and attitude of the participants was that statisticians
arrive at their priorities through the process of negotiation, by deciding
what the goals of government policy and users are, by determining how these
goals can be advanced through statistical information and the statistics needed, and by then negotiating with the respective parties to provide the requi-
site resources. The pragmatic situation, the bargaining situation, will be
with us forever. Thus, the present seminar did not seek a final mechanism for
setting priorities, what they hoped to achieve was a systematic procedure for
bargaining among users. The seminar would also provide a valuable forum for
sharing experiences in setting priorities and for studying approaches to
priority setting, approaches which might provide techniques which would make
the negotiation process more effective and efficient. Three approaches, a
framework for statistical planning as discussed in Dr. Duncan's paper, cost
benefit analysis, and the determination of priorities by who is willing to pay,
were discussed by the participants.

A Framework for Statistical Planning

In discussing the environment for statistical programs in the coming decade,
one of the main concerns had been the resources that would be available for
statistics. The potential value of a framework document, such as the one being developed in the U.S., is that it helps to substantiate, in a rational
way, the aggregate demand for statistics; it helps to support the resources for
statistics; it helps politicians see the overall value of statistical pro—
grams; and it presents a convincing argument for increased resources for statistical work. Two members of the U.S. delegation added that their agencies'
work in the Framework for Planning U.S. Federal Statistics had helped them to
share the views of their agencies' statistical programs, had forced them to
take a closer look at areas needing improvement, and had initiated an internal
review of their present and future statistical programs.

The European Economic Community's Statistical Office is now engaged in the
regular production of forward looking statistical programs; it is a "framework" in the sense that it tries to relate all aspects of their statistical
work within a single document and has much in common with the U.S. statistical
framework. Each of the programs discussed in the document covers a three-year
period with a forward look beyond that. There are background documents that

form the basis for decisions and also look at cross-cutting priorities. The document begins life as a discussion document, including a discussion for the policy makers. The document is then submitted to the council of ministers for their approval or disapproval of the general program (there is no budgetary or other approval for any single program). The document ends its course as a statement of priorities with the ministers general approval of the shape of the overall program.

The Conference of European Statisticians' (CES) "Program of Future Work" is also a framework for planning the work of this body. This framework has given them a balance and systematization in their program of work. Another idea in determining priorities, not mentioned by any of the countries, but used in the CES program of work, is "setting priorities in time." The Director of the ECE Statistical Division explained that if a certain program cannot be conducted in the first year, then they do it in the second year,...or the third, etc. Thus he suggested that setting priorities need not necessarily be a matter of "yes" or "no," but "yes, we do it this year," or "yes, we do it later." The setting of priorities for building up the infrastructure of national statistical offices might be of this nature. As it has to be done slowly, the question should be "what do we do first?"

Cost-Benefit Analysis

Some of the participants argued for a cost-benefit approach to the setting of statistical priorities, suggesting that priorities should be based on social cost and social benefits, including the direct cost of statistics and the indirect costs such as the cost to those providing the statistics. The cost of individual programs should be specified and related to social benefits.

Julius Shiskin, Commissioner of the Bureau of Labor Statistics (BLS) of the U.S., suggested that certain cost-benefit relations, although not explicitly derived, are very obvious. The benefit derived from some economic statistics can be very great compared to the amount in the statistical budget for these programs. For example, it could be very costly to a country's economy if an error in judging the imminence of a cyclical turning point led to an expansion policy when an expansion policy is no longer needed. And there are some cost-benefit relations that can be derived explicitly. For example, in the U.S. everytime the Consumer Price Index changes one point, more than a billion dollars is transferred from one group to another. Last year the unemployment figures were used to distribute about 10 billion dollars. In total BLS estimates that 25 to 30 billion dollars changes hands each year on the basis of statistics published by the BLS, whereas the budget for BLS is only 80 million.

However, other participants expressed grave doubts as to whether cost-benefit analysis would ever be an applicable technique for statistics. In the U.K. all they had been able to do in a cost-benefit approach was to measure how the absence of certain data would affect programs for the mid-term census. Another delegate cited a U.S. paper giving 25 examples of the use of cost-benefit analysis. His conclusion from the paper was that very little had come out of cost-benefit analysis.

Another observation was that statistics are only an intermediate product, the final product being the value to society as a result of having the information. Thus it is very difficult to view statistics as a final product in a cost-benefit approach.

Thus, the general consensus of the Seminar was that although cost-benefit
analysis cannot be used as a scheme for setting statistical priorities because
of the great difficulties in quantifying the cost and even more so in quantifying the benefits, statisticians should approach priority setting in the
spirit of cost-benefit analysis. Chief statistical officers should have a
sense of the cost involved, both the direct cost and the indirect burden, and
the benefits to society of having the information.

Charging for Statistics

Several countries mentioned their experience in charging the cost of statistical programs to the users of the statistics or of having one agency conduct
statistical programs for another on a reimbursable basis. The argument for
this approach is that "if someone is really willing to pay for a statistic,
then we can have a feeling that it is worthwhile."

However, determining statistical priorities by who is willing to pay for it
or by expenditures in the field may be misleading for several reasons. First,
for an individual field of statistics you may have to look at a lot of different fields of expenditure. Secondly, this approach puts the decisions in the
hands of those who have the power or money to pay for it and not necessarily
in the hands of those that use the statistics the most. For example, in the
U.S. the Council of Economic Advisers is one of the primary users of economic
data, but it is also one of the smallest agencies having only about 35 people.
If they were required to pay for their statistics, they would have to have
about a 1000 fold increase in their budget.

There are other problems in using this approach to determine priorities.
There are limits to the internal capacity of the central statistical office.
Offices may not have enough professional manpower even if someone is willing
to pay for it. Another difficulty relates to the situation in which statistical offices may build up a statistical capacity in response to the anticipation of certain persisting revenues from clients who are willing to pay for a
series over time. When the financial situation changes and the client dis-
appears, the statistical office is "left holding the bag." Also these surveys
may have become embedded in general use. If the original client cuts it off,
then there is a whole new clientele who wants it. Thus the statistical office
is forced to continue the program from its own resources. Lastly, there is
the problem of response burden; whether or not someone is willing to pay for a
series, does not take into account the additional response burden created.

The System of National Accounts

The Seminar also discussed the system of national accounts as a framework for
setting priorities among economic statistics programs. It was generally a-
greed that, although the national accounts is an important instrument in determining gaps, it is not a mechanism for setting priorities. They cannot
serve this purpose because most demands for economic statistics are for those
outside of the national accounts. However, the national accounts do have
other uses. First, because of the importance attached to the accounts, they
are used heavily in the decision process. Secondly, the analysis of the final

output of the accounts can identify discrepancies and needs in statistics. Thirdly, some surveys consistently over or under-estimate; it is easy to iden-tify these through the national accounts.

Thus, the answer to the question, how do we decide what to do?, is that we de-cide through negotiating and bargaining with the users of statistics to make certain that the efforts are relevant to the needs. Borrowing from the cost-benefit analysis approach and the charging for statistics approach, statisti-cians should negotiate in the spirit of cost-benefit analysis and with a user orientation in mind.

Several years ago one of the participants had discussed with his colleagues the different methods of setting priorities. Although they had approached their priorities in different methods, he was doubtful that there was any difference in the outcome. Thus he suggested that perhaps priority-setting approaches are very important in principle, but in practice they work out the same.

What Is Our Ability to Carry Through on What We Decide to Do?

But even if central statistical offices do have a fine system of setting pri-orities, will they have the power to implement it? Several participants ex-pressed concern over this aspect. One delegate said that his country had an elaborate machinery for setting priorities through committees. However, it involved so much work for the agencies and there was so much information that they couldn't digest it in time to make decisions. In the end, they gave up. Also, statistical offices are not autonomous, they have to react to govern-ment, to user groups, etc. Even if the statistical office has set its own priorities, there will inevitably be conflicts among the users.

The key to the negotiation process and to the key to having the ability to im-lement the process is for statisticians to enter the process from a point of strength in knowing what should be done in statistics so that the end result is a contribution to the total statistical system and to society. It is not enough to set priorities among statistical areas, priorities must also be set within the statistical system. That is, before beginning the negotiation pro-cess, chief statistical officers should decide whether high priority should be given to national or local government, to all the statistics necessary for completing the national accounts or to other accounts, to field of high poli-tical priority or of high government expenditure, to statistics of numerous use or to statistics from which we can derive a lot of information but do not have a high cost, to current statistics from which we can derive quick data or to long term programs. Also statisticians have to set priorities within a functional area, decide the balance between collection and analysis, determine the relative value of statistics versus a program area and evaluate new sta-tistical programs against the improvements in the timeliness and quality of existing statistical programs.

Another part of setting priorities is in terms of developing the appropriate infrastructure. Rationalization efforts of statistical offices must be accom-panied by the establishment of an appropriate infrastructure to respond more effectively to user demands.

Who Sets Priorities?

Who is responsible for setting priorities, for determining the intent of the
negotiation process, and for establishing a properly integrated statistical
infrastructure? The setting of priorities is not the statisticians' responsibility, nor the politicians', but a joint responsibility of the statisticians
and the politicians. However, it is the chief statistical officer's responsibility for determining the intent of the negotiation process, for seeing that
the statistical infrastructure accumulates over time and over functions, and
for being anticipatory about the needs of society to which he should be responding. The chief statistician cannot shirk this responsibility and it is a
very heavy responsibility if he is going to be anticipatory. Chief statisticians have to predict society and the role of government and then derive statistical programs from their predictions. Politicians do not look far ahead;
thus, the chief statistical officers have a responsibility to look a little in
the future.

Priority Setting in Countries with Centrally-Planning Economies

The discussion describing priority setting in countries with centrally planned
economies, as opposed to priorty setting activities in countries with market
economies, indicated that the statistics came out of a stronger planning context. These countries did not seem to have many of the problems identified in
Dr. Duncan's paper and by members of the Western European countries.

In the centrally-planned economy countries, the basis mechanism is the comprehensive state plan which yields a set of data needs. This is the force that
drives statistics. The planning process defines the themes for which statistics are needed. A working program is then developed to carry out the needs
for statistics. The priorities are confirmed when the plan is accepted. However, if unanimity cannot be reached, the chief statisticians can go to the
head of the Council of Ministers to resolve the issues. The final decision is
always made by the chairman of the central statistical system. The end result is a five-year plan for the central statistical system.

Thus, the experience of the countries with centrally-planned economies is
somewhat similar to the negotiation process which occurs in the market economy
countries and their process of meeting the needs of the various ministries for
information is not very different from that of the market economy countries
meeting the needs of the various programs which have been established by legislation.

Although the seminar did not come away with a tool-kit for setting priorities,
they agreed that the pressure was on them to rationalize the system as much
as possible and to build the system in a balanced way to meet the current and
future needs.

The Impact of Technical Development on Statistical Policy in the 1980's

L. Bosse* and J. Lamel**

*President, Central Statistical Office
**Secretary, Economic and Social Advisory Council, Federal Economic Chamber, Austria

The impact of technical development on statistical policy during the 1980's is likely to span a wide spectrum. This note deals with a few selected aspects which, in the Austrian Central Statistical Office, are clearly felt to become significant.

Coordination of Information Concepts

As a consequence of general computer usage and growing software power, it is foreseeable, if not already experienced, that a variety of information systems and data banks will successively be established in the different areas of public administration activity. These systems, designed to meet the specific administrative requirements imposed, may allow for communication with selected other systems in order to facilitate local co-operation. But they will rarely be based on detailed considerations concerning the general information compatibility problem we will be faced with in the future.

When talking about compatibility in this context, it is with reference neither to hardware nor to software problems, but to the compatibility of information space concepts which inevitably must be established whenever an information system is planned, no matter what type of data base structure and management system is chosen.

There are a number of reasons why the above statement on lacking compatibility considerations in the said sense will, in general, be true. First, to a specific administrative agency, there is no direct need for attacking the problem. Secondly, there is not enough information on information to appreciate its importance. Finally, the prospects of engaging in it are not at all appealing.

Nevertheless, the challenge of growing systems with - more or less - isolated information spaces calls for preventive action because, as a general rule, the more complex systems grow, the more costly - if possible at all - will be the tackling of compatibility problems which were not solved in statu nascendi.

If the functions of a central statistical office are to be understood as those of a national (statistical) information service, it is conclusively charged with the care for co-ordination of information space design work in

governmental agencies. It will probably become one of the primary lines of
statistical policy - in order to meet this task - to adopt adequate long term
measures which, in concreto, will of course highly depend on the legal and
structural milieu in which the national statistical office is embedded.

Intra-Agency Organization

Direct and secondary consequences of computer usage have already repeatedly
provoked the question of whether the traditional organizational structure of
the majority of statistical agencies (which is of the subject matter type)
will still be adequate in future times.

It is firmly believed that this question will grow more demanding in the 1980s
rather than cease. The technical and methodological development of the past
decades repeatedly led to the extraction of specific functions, till then
embedded in substance matter divisions. Special organizational units were
created and charged with those specific tasks for efficiency reasons or as a
necessary environment for the use of highly specialized new tools. This process, which is seen as an act of centralization from the traditional viewpoint
of substance matter structure, is clearly not yet terminated. In view of
forthcoming integrated systems and information systems, for example, a new
responsibility of data administration will have to be established sooner or
later. It is hardly imaginable that this competence be split and allocated to
various subject matter groups.

Moreover, the production of statistical information in these decades is taking
on a somewhat industrial scale which, in turn, asks for new functions (like
priority scheduling, calculatory work, etc.) which organizationally must be
allocated in some way but, by their nature, must remain outside of proper
subject matter activities.

It may be seriously felt that this - continuing - evolution will drive the
primary substance matter structures out of balance. On the other hand, the
original reasons for setting up subject matter main structures lay in the
existence of not too big a number of clear-cut areas of statistical work.
This is no more reality. Statistical activity nowadays is spreading over
dozens of "subject matter" fields and - more important - is faced with increasing demand of intersectorial analysis.

There are national statistical agencies that have already taken action towards
a more adequate internal structure, but the majority, as yet, has not. The
problem, therefore, will perhaps become one of the most important - and most
difficult - objects of intra-agency policy in the 1980s.

Information Distribution

The mode of information supply to the consumers of national statistical services has already remarkably changed during the last decade.

While, in former days, the distribution of statistical information was almost
totally confined to printed publications, recent strategies tend to establish
at least one additional level, i.e. the preparation of work tables, not
published and in most cases not even printed, but stored on magnetic media.

With the help of generalized programmes, tables or subtables are retrieved from these data stocks at request. The consumer is, if at all, only charged with the cost of machine running time which in general is modest if the programme operations involved are rather of the selecting than aggregating type.

Statistical macro data banks being developed or planned in these days, particularly if equipped with on-line facilities and user-oriented retrieval languages, will presumingly make such data stocks so attractive for the user that they may considerably affect the domain of the traditional printed publication.

Future information distribution policy will probably have to conceive statistical data banks as a specific means of publication. The printed brochure, in the future, may well be freed from the load of monstrous, space-consuming tables and, instead, focus on well-perceptable presentations of significant results of general interest and, in addition, act as a guide concerning scope and availability of more detailed information.

It can be expected that, in consequence of the information systems development as a whole, there will be a general shift towards the on-request-supply policy. In view of the diverging demand profile on the consumers part this tendency will probably be well in line with economic considerations.

E

Anticipated Computer Hardware and Software Developments and Their Relevance to Statistical Offices

I. P. Fellegi* and E. Outrata**

*Assistant Chief Statistician, Statistics Canada
**Statistics Canada

INTRODUCTION

Changes in the nature, extent and mode of operation of statistical offices
could occur due to any of the following five major impacts: the changing
nature of data needs of users, changes in sources of data supply, budgetary
circumstances,technological developments and methodological developments.

The single largest challenge facing national statistical offices concerns the
increasing data needs of more rational decision-making at all levels: international, national, regional, corporate and personal.

Therefore, the statistical office must rely on rational planning. Increasingly, planning will also have to consider the impact of decisions on the
evolving environment, as affected by a variety of economic, social and political factors. Thus, it is very likely that planning will be increasingly based
on sophisticated quantitative analysis.

In order to support anticipated decision-making processes with relevant data,
statistical offices will require means to adapt to changes as they happen in
the structure and quality of these processes. Some trends can be readily
seen from today's vantage point, like the need for attitudinal and perceptual
data or measures of causal relationships; the need for time series and integration; small area detail.

The need to ensure that data available potentially in the statistical system
are practically accessible to the decision making process is reflected in
growing demands for fast response, for data on the quality of information,
and for data on data, which inform users about what information exists and
how it can be accessed.

On the side of data supply, important changes are expected in increased
public resistance to surveys combined with real or imagined concerns about
privacy and growing threats to security. On the other hand, wider use of
administrative files, particularly for calibration, and automation of files
of large businesses and institutions, is expected.

This environment, together with a scarcity of funds that will require systematic reevaluation of existing programmes, will lead statistical offices to a

mode of operation that stresses separation of the collection of data (which
will be regular and will use as much as possible information that does build
up problems with respondent burden, etc.) from the use of data for publication
and analysis (which will take place at the time of its need). A new dimension
will be added in the need for objective descriptions of data quality and for
full accessibility.

The following text refers to the anticipated technological and methodological
trends that are relevant to and concurrent with these developments.

TECHNOLOGICAL DEVELOPMENTS

Hardware Developments

When considering the hardware development predicted for the next 10-15 years,
the decrease in cost and size of the hardware components is the most specta-
cular feature.

Improvements can be expected in the cost per unit of processing, cost per unit
of storage and cost per unit of transfer. Speeds will continue to increase.
Where basic switching and transfer speeds are reaching physical limits, increasing use will be made of parallelism.

Central Processors and Architecture

The main line of central processors will continue to offer improved price/
performance at the rate of about 15% per annum. Each manufacturer will provide, in addition to price/performance improvements, enrichments of the functional capability of new systems. These enrichments will be in the areas of:
useability and reliability, availability, serviceability/RAS/.

Useability features have been added to systems in the past by enhanced expand-
ability, connectability of more and larger units, debugging aids and other
features. Because the cost of maintenance continues to rise while the cost
of circuitry goes down, more hardware features to assure reliability will be
added. These will be largely transparent to the user.

The performance of large-scale computers has already been significantly
enhanced by increased parallelism in software (multiprogramming) and hardware
(multiprocessing). They will more often than not have large numbers of terminals. At the low end, minicomputers in a variety of uses and applications
continue to gain acceptance. They will increasingly be incorporated in remote
terminals, key entry systems, peripheral controllers, and communications con-
centrators. These special function mini computer systems will tend to be
combined in one system in the future, and will evolve into remote data col-
lection/management subsystems.

As the number of real time applications increases, multiprocessor systems will
become more frequent. Multiprocessor systems consist of two or more central
processing units (CPU) that are specially synchronized and linked through a
common shared memory. Their main advantage is in the vastly increased reliability of the system because of the redundancy of processors. This ensures

that if one of the two processor experiences a "down," the other one automatically takes over, thus providing an uninterrupted service to users.

Finally, more and more computer networks will evolve. A network consists of a number of computer systems interconnected by communication lines. Networks can be organized in various ways. At one extreme each computer system in the network can be independent and the lines are then used merely to send messages from one computer to another. If the lines were not used, the messages could be sent by cards or tape. At the other extreme is a network organized as a hierarchy with one master system which "controls" a number of other systems, these in turn control some other computers and so on. In between these extremes are numerous forms of cooperation and control.

The primary result of these developments will be an increase in flexibility and enormous increases in the alternatives available to the designer, particularly in multilevel integrated systems in which small hardware systems can be used to satisfy local or specialized functional needs (e.g., data entry and on-line primary editing) and still be integrated with larger central units which will maintain and operate on the overall corporate data base.

Storage Hierarchy

Storage technology is expected to develop along two main lines. Firstly, the capacity and speed of the conventional storage media such as tapes, discs and drums will increase notably. Secondly, new techniques such as "bubble memories" and holographic storage media will come into use. The statistical services have a need for very large archival memories with low cost and, depending on the application, fast access, since the volume of data that they would like to store in the future is enormous due to the fact that, theoretically, all previously collected data remain of potential interest to the statistical system. The need for fast access is determined by the particular application, i.e., the extent to which increased speeds can be taken advantage of for the improvement of the total man-machine systems. The primary objective is to be able to retrieve large and arbitrarily defined sets of data from the archives and generate tabulations from, or apply analytical techniques to these data. To make this possible it is necessary to have short access times to the individual data items. An ideal solution for the statistical service would be extremely large content-addressable stores at low cost.

The increasing demands for rapid access to storage will be met by many products, at reduced cost, with increases in speed and storage capacity. The concept of virtual storage will become an important hardware and software feature in the future. Virtual storage is a software development which allows users to write application programs as if there were no physical storage constraints in the main memory of the computer--the actual physical storage limitations are taken care of by the software which swaps data into and out of the actual physical memory available as needed. Thus as far as the programmer is concerned, he can write his programs as if the memory was infinite and as if all the data needed by his programs were resident in it at the same time. This can lead to significant simplification of program structures, but at the price of increased hardware utilization.

Data Capture

Two important trends in technological development will have a marked influence
on data capture operations. One is the development of data communications and
terminals. This will lead to much more physically decentralized data capture
than today. The development of microcomputers which can be easily attached to
data preparation equipment also makes it possible to attain a greater degree
of input editing at the data entry stage. The second important trend is
source data capture. This means that source data is automatically recorded in
conjunction with an operative process: e.g. by the accounting and MIS processes
of business respondents or through small portable keyboards with microcomputers
and tape cassettes or discettes for data entry (and some verification and/or
prompting) "on the spot" by interviewers.

An alternative to the approach outlined above (i.e. automatic data capture as
part of the operative process generating the data) involves the collection
of data in a form which subsequently can be converted into machine readable
files with minimum manual intervention. Optical Mark Reading Devices (OMR)
and Optical Character Reading Devices (OCR) are currently in use in the bureau.
Mark readers were used in the 1956, 1961, and 1966 censuses and also the
Labour Force Survey. FOSDIC also is a special kind of mark reader. Within
the last year an IBM/1288 OCR reader has been installed.

OCR systems can be utilized for data entry by using a type and scan technique.
In many instances this approach has proven efficient, generally outperforming
key to tape/disc devices. A major improvement has been the addition of the
signature capture feature which allows a video image of unrecognizable characters to be displayed by an on-line CRT for correction. This feature has great
possibilities for automatic, or on-line computer assisted coding of write-in
items such as occupation and industry.

Data Presentation

If developments in new data capture methodology have lagged in the past, the
situation for data dissemination has been more so. The amount of printing
done each year is increasing and now exceeds a billion lines. Excluding
census, the bureau produces some 100,000 original pages of publications each
year. Preparation of publications has been automated to some degree by the
use of ATS to store and update material to be published and use of photo-
composition to prepare copy. Significant improvements in timeliness and savings in costs have been achieved. However, the fact still remains that most
data is disseminated in the form of hard copy.

Developments within the area of data presentation clearly indicate a rapid
improvement in the possibilities of presenting more data in a way better
suited to the user than today. The main features will be increased interaction through terminals and the increased use of machine readable and graphical
output.

In the area of visual display terminals, major breakthroughs are expected in
the 1980's, many of which are already apparent. The cathode-ray tubes presently in use will be replaced by displays which are much more flexible.
There will be improved possibilities for graphical and colour display at lower
cost.

With respect to photo-composition, new developments are underway which will render the production of visually pleasing printed publications much more accessible, both from the point of view of economies and ease of use. The technology is already in place for a much broader utilization of machine readable output relative to hand copy.

Data Communications

The government's telecommunications policy is designed to ensure that tele-communication facilities be developed in a competitive environment, but consistent with some national priorities among which, for purposes of the present report, the most relevant is that telecommunication charges to clients should not penalize users located in relatively remote areas. It is safe to forecast that telecommunication costs per volume of data transmitted will be significantly reduced, irrespective of the geographic location of users.

Data communication can be used in the following cases:

Remote job entry. A batch job for the central computer is initiated from a terminal some distance away. The output of the job is annually dispatched back to the terminal. For example, our Regional Offices could in the future submit some batch jobs for processing in the computer centre in Ottawa.

Time sharing. In such a system many people use one computer simultaneously in an interactive mode from remote terminals.

Enquire/response. Queries are entered from a terminal to a data base management system. Response is in real time to the terminal.

Collection/distribution. Decentralized data entry and distribution of reports.

Data communication by personnel who are employees of Statistics Canada poses no major problems. In fact, within the complex of buildings presently occupied by Statistics Canada there currently exists an extensive and physically quite secure data communication network. Moreover, our regional offices are also presently connected to Ottawa through secure data communication lines. What is forecast is that the cost of using such a network will be reduced significantly, rendering decentralized data collection and retrieval more attractive. Bureau access to physically decentralized administrative files, both federal and provincial, will become economical from the point of view of communication costs--provided other prerequisites are resolved.

From the point of view of our clients, the same technological developments will render their on-line access to our data bases equally attractive. Present technology with respect to data base security and, more importantly, the state of the art of statistical disclosure analysis, rules out for the present time this type of access to our micro-data files. However, the technological problems of improved data base security will undoubtedly be solved within the next few years and the methodological problems of statistical disclosure analysis also show promise of being resolved in the near future. Thus the present barrier of remote access to our micro-data files for purposes of retrieving statistical aggregates or analyses will likely be removed. Of course, remote access to aggregate data (of the CANSIM type) is a current reality whose computer/communication costs are likely to decrease over time.

In order to simulate some of the problems of remote user access to micro-data
files (for purposes of retrieving statistical information), we could currently
experiment with "soft" access to microdata files, as described elsewhere in
the present report.

SOFTWARE

Introduction

Since the beginning of computing in the late 1940's, the relative importance
of hardware and software in cost, reliability and performance has moved continuously from hardware (whose cost was overwhelming in the 40's and 50's)
to software (whose cost is estimated to become 80 to 90% in the early 1980's).
This is due to the methodological difference of the two divisions of computing:
hardware development has from the very start been an engineering discipline
based on scientific development and has maintained this characteristic throughout, while software has developed from its beginning as a craft and, despite
some limited success in marginal fields, has to this day not been able to
develop a comprehensive theory and a scientific foundation. Consequently,
partially for this reason and partially for causes specific to the nature and
subject-matter of the discipline, for a longtime hardware development has
managed to maintain extremely high rates of productivity growth. Software
development, on the other hand, has developed like a craft: by implementing a
few techniques which experience has recognized as being advantageous. The
hardware production shop today is a factory; the programming and systems development shop can still be considered in analogy to an artisan's studio.

This situation is not likely to change between now and the middle 1980's,
because no development of a theory of computing is even on the horizon as far
as we know. (Some significant exceptions will be mentioned in subsequent
paragraphs, but these do not form any common pattern of knowledge). Even if
the foundation of such a theory were being formulated at this moment, it is
still unreasonable to expect any effects of its acceptance, development and
complete implementation in practice within the next 10 years.

The major concern for system implementation is, therefore, a search for
methods that would minimize the constantly growing costs of system development,
assuming basically the present state of the art. The expected attack on the
cost of system development will therefore be led not frontally like in hardware (not by radical revolution in construction and techniques), but by con-
tainment through minimization of the amount of work that has to be done,
through automation where possible, and through improvement and rationalization
of project control.

This having been said, it should be emphasized on the other side that the introduction of data base management systems suggests change in at least some
important characteristics of systems work. To what extent the introduction to
data bases over the next 10 years will actually lead to a revolution in system
design is yet to be seen; at present the areas of both theory and practice are
still rather in their formative phases and the long term results of implemen-
tations are as yet unknown. A statistical office, however, conceived as a
national centre for providing information, is an ideal data base application,
at least in theory; and, therefore, it is also a prime challenge for this
method.

General Trends in System Development

It follows from the introduction above that the main way to increase reliability and cut costs of newly developed systems is to minimize the amount of software that has to be developed; and where this cannot be avoided, to establish an environment in which this development can be performed in the most controlled manner. This cannot solve the "software problem" but can help to reduce its effect.

Use of hardware. The boundary between hardware and software is somewhat artificial: it is always theoretically, and sometimes practically, possible to construct hardware to perform the function that is needed. While the extreme solution of building special machines to perform the function of individual systems is probably not applicable to most areas of computing in a statistical office, there are certain fields in which acquisition of hardware devices may reduce or eliminate the need for software development, or at least for the maintenance of special software. As has been suggested above, hardware costs fall much more rapidly than those of software, and this suggests the need to evaluate often and repetitively instances where decisions in the past have been made not to acquire special hardware, but to use software operating on general hardware. Prime candidates might be APL, and more generally, interactive computing. Other instances will undoubtedly be added over the decade.

Minimization of new development. Much more emphasis will be placed in the future on the re-use of parts of existing systems during the development of new ones. In the profession as a whole, and most certainly in Statistics Canada, the present practice of system development tends to propose much too often complete development of all parts each time a new system is designed. It can be expected that during the coming decade, both users and system developers will attempt to construct systems so that their parts are re-usable in other applications. It must be remembered, however, that effective application of this principle is an element that surpasses the computer system alone; it implies not only specific techniques in the construction of systems (the top-down approach) and better knowledge of existing re-usable parts of other systems on the part of system developers, but also a higher level of standardization and integration on the part of statistical methodology and users in general. This approach in both the system area and in general is at present only in its infancy; and progress in the past has been limited to isolated islands (e.g. tabulation, certain areas of editing, file manipulation, etc.).

Automation of new development. One way to minimize the need for new development is to construct translators of very high level languages specialized for particular functions. This approach is similar to the one above, in that it requires less new development. It differs from it in that it introduces a new level of software (which is general to many applications), and reduces the amount of development work by facilitating the way systems are expressed. Such development is again still in its infancy, while its need has been recognized a long time ago. Examples of its application in Statistics Canada are GEISHA and some tabulating systems. The ultimate would be a language for system development where parts of systems would be expressed in a simple non-procedural way; some such ideas are being discussed at present and should impact the field of software development during the next decade. Again, the statistical office is in an especially good position to implement such developments, because of the fundamental repetitiveness of its work.

F

Improvement in project control and methods of measurement. All that was said
in point 3.1 notwithstanding, it is yet true that even in the present state of
the art of computer programming and system development, it is possible to
introduce a significant amount of measurement that should make estimating and
control of projects more rational and precise, and results in costs and reliability more predictable, We should expect in the next decade the introduction
of effective methods of management and control of development. Where special
systems must be developed, such methods should allow their smooth management
and organize the exchange of information necessary for improvement of methods
used in the craft. We may, therefore, expect some improvement in productivity, particularly an improved ability to avoid catastrophes; this improvement,
however, will not be anywhere close to that demonstrated by hardware development in the past.

Impact on Statistics Canada

Efficiency and reliability. The general trends in the system development
profession discussed above and greater awareness of them among users of data
processing in Statistics Canada will help to introduce new measures and practices in Statistics Canada. The following changes and developments may be
expected:

Improvement of estimating processes. Present practice of project development
is only beginning to rely on a systematic methodology for the estimation of
system development cost. Introduction of mechanisms that will improve the
ability of developers to estimate can be expected.

Improvements in the planning and organization of projects. Full implementation of accepted methods of planning, organization and management of projects
may be expected and should be facilitated in Statistics Canada in the first
years of the decade.

CLOSER INTEGRATION OF DIFFERENT ASPECTS OF DESIGN AND DEVELOPMENT

Subject-matter, methodological, and system design in a project are three
different aspects of the same process. Therefore, a better understanding may
be expected in the coming decade of the fact that optimization of the total
system (development or production) is always possible only when all aspects
are taken simultaneously into account. This understanding should lead to
closer integration of the subject-matter, methodological, and system components in the design and development process.

EVALUATION

A greater understanding of the need for subsequent regular evaluation from all
aspects of existing production systems will in the next decade lead to the
introduction of mechanisms that will ensure that it is done.

APPLICATION OF NEW SYSTEM DEVELOPMENT AND PROGRAMMING METHODS

Modern programming methods, like structure programming or the top/down

approach, are being implemented now. The result, among other effects, will be
increased modularity in our systems--which, in turn, should lead to improved
reliability of systems as well as a greater potential for re-using parts of
previous systems when designing new ones. Organizational structures and
mechanisms will be developed, which will allow the early introduction of other
improvements in programming that may arise.

Current work on a theoretical basis for the development of automatic methods
for proving the correctness of programs will probably not have practical
results on the efficiency and reliability of system development within the
next 10 years. Thus, we are limited to development and implementation of
testing strategies within the cycle of systems development as the only practical method of assuring the adequacy of algorithms and programs.

LIBRARIES OF MODULES

Development along the trends described above requires some effort in the
establishment and institutionalization of the exchange of information about
existing and developing program modules in a structured programming environment. In the 10 years ahead of us, this should be solved and well entrenched.

Data Independence

A software product which does not need to be altered when the data to which it
refers undergoes a change in structure can be said to be data independent.

The most promising development in data independence is the introduction of
data administration and data management systems, which allow multi-purpose use
of centrally managed data. The user programs interface with the data through
the data management system, which handles all physical storage. Thus application programs are relieved of the need to attend to physical manipulation,
thereby reducing the developmental task. Fundamental improvements in designer
productivity and system reliability can result from the concept of data
independence for two reasons: application programs can concentrate on the
logic of the application to the total exclusion of worrying about physical
data handling (which currently accounts for a large proportion of program code
and which "clutters up" program logic); and because changes in the data structure (such as e.g., a new questionnaire for an annual survey) need not result
in extensive program changes so long as the substantive data collected previously is still available after the changes.

A basic prerequisite for the introduction of data independence is what is
known as a Data Base Management System. This concept is described in the
next section.

DATA BASE MANAGEMENT SYSTEMS

Data base management systems are systems placed between operational software
and application programs, which ensure access to a randomly organized data
base from a variety of programs that are generally linked to each other by
nothing more than the knowledge of the properties of data base management
systems.

Data base technology has developed slowly from the beginning of the '70's and has now reached a point where different concepts, implemented in a dozen or two commercially available packages, compete for partial or general recognition. The long term importance of data base management systems follows primarily from their potential to radically change the conceptual approach to system development.

Traditionally a computer system is developed as a set of programs, each performing particular functions, in a specified sequence. Each program uses one or more files of data and produces another file. By subsequent processing of intermediary files, the required data are eventually derived. The efficiency of program execution often requires that the intermediate files be processed in a particular sequence--that sequence not necessarily coinciding with the sequence which would be dictated by the logic of the problem itself.

A data base management system can potentially revolutionize system design by allowing the designer to conceive of a system as a set of operations performed relatively independently on the data that are required for these functions. In other words, the designer may disregard the need for establishing, organizing, and maintaining data structures and formats, as described above under data independence. In final analysis (again in principle) this allows the designer to split the system into a series of independent functions sequenced only where logic of the task itself demands sequence.

Being the collector and provider of information par excellence, Statistics Canada is in a unique position as concerns data bases: in concept it is a large data base. Today it is probably limited by the state of technology and its size and complexity from being implemented in that way, but this interesting quality makes it a prime user of new developments in this area. It is probably, therefore, that during the next decade data base management will be introduced on a large scale in Statistics Canada. Because this prospect could be the most radical change in system development in the next decade, we elaborate on this concept.

DBM systems represent major steps toward the simplification of application programs because they are capable of cross-referencing the internal (physical storage) representation of data with the external representation of data (i.e., the form in which application programs would like the data to be). Thus, for example, DBM systems free application programs from the need to refer to the specific physical location of records and parts of records-- instead they can refer to data elements through their suitable economic (English-like) names. This goes a long way toward achieving data independence as outlined above.

In spite of the major contribution of existing DBM systems to data independence and simplification of application programs, the currently available versions of such systems are still subject to major problems. They will allow a wide variety of physical storage formats, the data base design itself is more of an art than science, they do not achieve complete data independence, depending on the data structures and the amount of redundancy in the data they are subject to error-prone updates and they tend to be burdened (through their very flexibility) with high operating overhead costs.

So long as DBM systems try to accommodate the infinite variety of possible data structures (which, almost naturally, commercially available systems attempt to do), the shortcomings indicated in the previous paragraph are unlikely to be overcome. What is required is a theory of data structures which could lead a data base designer in an unambiguous fashion from the nature of the data to be stored to the design of the data base. By analogy, so long as data editing was conceived as an ad hoc operation of diagnosing a variety of exceptional situations (errors), edit programs were always difficult, hand-crafted and error-prone. The recently formulated theories of editing lead in a more or less unambiguous fashion from the concept of error conditions to their representation as edit statements. Moreover, the theory proves that all edit statements can be expressed in a very simple and uniform manner. Similarly, a recent theoretical development appears to be of fundamental importance with respect to data structures and, indirectly, with respect to DBM systems. It can be proven that a file containing any arbitrary data structure can be broken down in an unambiguous fashion into a series of subfiles, each subfile having an essentially identical structure; consisting of a series of records each of which is of fixed length. Thus any file, no matter how complex its data structure, can be broken down into a series of subfiles of a uniform and very simple kind.

This development, called the relational model, seems to hold the key to the standardization of stored data structures, as well as for the construction of Data Base Management Systems which are particularly effective for statistical data processing. Combined with standardization of concepts and nomenclature (a major prerequisite!), as well as parallel developments in computer networks, this development holds out the opportunity for the development of distributed data bases (physically and even organizationally) and thus for the evolution of a truly national statistical information system encompassing all relevant data sources, and including appropriate provincial and federal administrative files with statistical content.

Possible Trends in Using Technical Means for the Further Improvement of Accountancy and Statistics in the GDR in the Decade 1981-1990

K. Newmann

Head of Research Center, Central Statistical Office, German Democratic Republic

The Use of Statistics in Planning

The demands on governmental statistics in the GDR will be determined, also in the years beyond 1980, by the needs of all the socialist society to be informed of the socio-economic processes that have passed, and of their effects, including those to be expected in the future, in a universal, consistent, exact, and timely way. Thus, statistics contribute to the implementation of the long-term concepts with regard to the systematic improvement of management and planning in the national economy. Both quantitative and qualitative changes, which are presently occurring in our national economy, characterized by increasing intensification of the entire reproduction process under the permanent influence exerted by scientific and technological progress, are placing new, much higher demands with respect to the central planning. Consequently, governmental statistics will have to fulfill its main function even more throughly: to collect data on the fulfillment of the long-term, medium-term and short-term plans at all levels of the national economy by branch and territory.

This requires taking thoroughly co-ordinated measures to ensure permanent improvement with respect to the contents, the methods, the organization, and the technological basis of accountancy and statistics. The experience gathered in the past development says very clearly that the requirements with regard to the contents have the greatest influence. Methods, organization and the technological basis always play a serving part, however active this part may be, as compared with the contents of accountancy and statistics.

The Technological Level of Data Processing

Apart from social requirements, the level that governmental statistics itself will have reached in the early eighties will strongly determine further development. It can be taken for granted that the integrated system of accountancy and statisics, which developed step by step in the sixties in the GDR, has proved successful, and that it has been and will be improved in the seventies in line with the improvement of all management and planning so that it will provide a solid basis for further developments. Gradually, it will

meet the criteria which the socialist countries united together in the CMEA
have taken as a basis for their future automated systems of governmental statistics. As regards the GDR, this process will develop in the course of the
present five-year plan (1976-1980) in a way that the experience gathered so
far in using a data bank for statistical purposes will be generalized so as
to gradually create the conditions for a data bank system of statistics. This
is done, above all, with a view to reaching significant improvements with
regard to the evaluation of the big stocks of data available without having
excessively to increase public outlays for their collection, transportation,
processing and storing. It should be stressed that the data bank system of
statistics will not yet have developed to perfection by the end of the
seventies. Thus, the presently prevailing form of the organization of data
processing by type of reports - for the various subject-matter areas and,
within them, for the periods concerned - will exist side by side with the data
bank system, being used as a data source for the latter one. It can also be
expected that operational data banks will be available at that time only for
central headquarters and for about one out of three regional statistical offices.

The Technological Level of the Environment in Which Statistical Offices Operate

However, it is not only the technological level of data processing within the
statistical services that determines the further development, but also the
technological level of their environment. To the extent as those who supply
the statistical services with data as well as those who receive information
from them, are equipped with technical means of their own for the processing
of information or make use of such means, the conditions of work for governmental statistics itself will undergo a fundamental change.

The difficult problem will be the fact that this process will develop evenly
in the next 10 - 15 years. This makes the organization of data processing
more complicated and also influences the demands on the technological basis of
statistics. Thus, it has to be ensured that the reporting units which are
able to furnish the statistical office with machine-readable data will be
allowed (in legal terms) and given the possibility (from the technical aspect
of further processing) to do so. Precisely the same is true of the distribution of the statistical results among final users. Concretely, this means
that the preference (necessary in the past and implemented at considerable
expense) given to standarized organizational means for the collection and
transfer of statistical information will have to be adjusted to the new
conditions. In this connexion it is necessary under all circumstances to
ensure the uniform nature of the contents and of the methodology of statistical indicators.

In the GDR the definitions of all indicators used in the field of planning,
finance and statistics are laid down in legal provisions, i.e. they have the
character of national standards. The same is true of all nomenclatures. This
ensures a high degree of standardization which makes it also possible that the
information is coded already in the reporting units. In accordance with the
orientation given in the five-year plan 1976-1980 with respect to the use of
EDP in the GDR, a great number of enterprises and institutions will automatically process retrospective data by 1980.

Since this scientific and technological development is organized, i.e. planned consciously in our society, it is also possible to use it consciously.

In this respect the social ownership of the means of production provides the basis for the overall utilization of statistical information, on the one hand, and a homogeneous technical policy approach to the application of data processing techniques, on the other hand.

The Utilization of Technology for Statistical Purposes

The demands on governmental statistics, the level of development reached by the statistical information system in the early eighties, and shaping of its environment it terms of data processing facilities are substantial conditions for the future utilization of technology aimed at the further improvement of statistics.

As a matter of course, the supply and the real availability of the technical means themselves play an important role. Once again it should, however, be stressed that the strongest impulses for the application of these technical means will go out from the contents of statistics and from the methods of processing, and that the volume and the type of these uses will essentially be determined by the way in which these evermore complicated procedures of data processing will be mastered from the organizational point of view.

Here we proceed from the well proved principle that we should not strive for a maximum of statistical information, but for an optimum. The criterion of optimality, also in the future, should be the usefulness of the information concerned for the improvement of management and planning which should be rationally related to the expenditures involved for achieving it.

Of course, our resources which the society can use to satisfy their needs for data, including the needs for statistical information, will be restricted also in the future. It can, however, be expected that the role of man in the field of data processing will tend to increase in the future, when automation will have liberated him from recurrent routine work. Further it can be expected that a tendency that can be observed even at present will continue: the relation between capacities and needs in the field of EDP will be highly differentiated both in terms of quality and in terms of time. The experience shows that the qualitative and quantitative increase in capacity--which is in general linked to the introduction of new EDP equipment--gives birth to the impression as though sufficient capacities are new available to cover the needs for information. However, since the needs for statistical information are a--socially determined--extremely dynamic value and also because increasing skill in using the available new hardware and software initiates, in its turn, new needs, the limitations of the available capacities will become visible again after some time, which can only be overcome by a new increase in the technological pattern (new hardware and software).

We suppose that in the GDR:

> At the beginning of the eighties a technical reequipment will just have been started which will be finished by the middle of the decade,

> The programming of statistical tasks will have reached a substantially higher level as is the case today with the direct use of general program languages (such as COBOL or PL/1); these new methods and techniques of

programming bringing about a decisive improvement in the communication
between users and computers,

The development of the technological pattern throughout the decade will
be determined most strongly by an on-going process of increasing utilization of the advantages of data remote processing,

In the last third of the decade a new technical reequipment will begin,
which, however, will become effective for the general organization of
statistics only in the nineties.

Data remote processing will lead to much more promptness in the use of statistical information. Together with the development of data banks, this approach
will make it possible that large circles of the society will easily and directly gain access to timely statistical information. Data remote processing
will also lead, in conjunction with data remote transfer, to machine-links
between computers installed in different computing centres, for instance, in
order to improve the exchange of data. Provided that enough experience will
have gathered and in accordance with the developing needs, at first provi-
sional and then permanent computer networks can be supposed to develop.
Branch and territorial computing centres for collective use will be the nodes
in this network.

A Data Bank System of Statistics

In order to make as full use as possible of the advantages that result from
these trends of development especially for statistics, it is necessary to step
by step build up a data bank system of statistics

The system will consist of the central statistical data bank, and statistical
data banks for the 15 counties (Bezirke) of the GDR (which can be at the same
time component parts of larger territorial data banks).

In its further development, it will become possible to establish links between
enterprise-owned, branch and other data banks if they include information
which is important for statistics.

Such system can only be operational if an efficient information system (i.e. a
meta information system) is used for its organization management. (According
to the present ideas, a data base management system might be considered the
embryonic stage for such system in the future.) A data bank management system
can be developed only gradually. At the beginning it has to rely on the
classic organizational means that have already proved their usefulness in
statistics such as e.g. information lists, catalogues, register etc. Systematic improvement and - if appropriate - automation of these organizational
instruments will provide the basic conditions for systems of retrieval and in-
quiry. At first, these systems will give a (possibly timely) survey of the
given state of the statistical information system (passive phase) and will
later open up the possibilities to change this state and, in this way, to
adjust the statistical information system to the developing tasks (active
phase).

Requirements for a Data Bank System

First of all the organization of a data bank management system will rely on
the active interrelation between the contents (of statistical information) and
the technology of data processing, which has in its turn a strengthening backward effect (feed back). The primacy of the contents requires that in this
process the logical relations between the various indicators be defined so
that they are in accordance as much as possible with the real relations between real objects (which are reflected by statistical indicators) (infolo-
gical approach). Recording of such relations in the form of (possibly "open"
or "dynamic") models which are subject to permanent updating is feasible in a
rational way only if efficient EDP technology (dialogue systems) is available.

The Effects of Technological Development

The development and the use of a data bank system for statistical purposes
will not only put high demands of EDP hardware and software in the more narrow
sense (central units, external storages), but it will also considerably influence the technological pattern of the preoperational and postoperational
stages of work.

Statistical practice will change above all as a function of the new possibilities to cope with data collection problems. It can be expected that data
inputs into the statistical information system will be made in the way already
mentioned by taking over machine-readable data carriers or - where this is not
possible - by using peripheral devices which, through data remote processing
media, are simultaneously linked up with highly effective computers (intelligent terminals). This form of terminal will make it possible to a certain
degree to do some primary data processing already at the lowest levels (in the
GDR: district (Kreis)), taking account as largely as possible of the information requirements of local government. As previously, the more pretentious
tasks of processing statistical information, however, will be carried out
either at the next higher level (in the GDR: county (Bezirk)) (which will have
powerfull data processing centres for collective use) or at the "Computing
Centre of Statistics" in a centralized way for the total republic. Statistical practice will also change by the fact that data output will reach a higher
technological level as well. Here, too, the presentation of machine-readable
data carriers will be supplemented by such techniques as, for instance, microfilms, photographic setting and the like.

All these and other effects of the technological development will increase the
efficiency of governmental statistics to the extent as we will be able to cope
with them in terms of organization and management and thus to devote them to
issue of improving the contents of statistics.

In this way the tendency that the technological development supports governmental statistics in its efforts to provide all the socialist society and,
above all, its leading authorities ever more completely with statistical in-
fromation will continue. At the same time it will expand the possibilities
for those who provide governmental statistical agencies with information to do
so with a higher quality and in a more timely manner. On the other hand, the
final users of statistical information will increasingly have auxiliary technical means which will enable them to use statistical data more fully and more
profoundly. That is why it can be expected that, apart from growing needs for
statistical information (in number and texts), a need for statistical approach-

es that can be reused under different circumstances will also increasingly develop.

All this, taken together, will further raise the role of statistics in the society.

Some Characteristics of the Technical Development in Statistical Organizations

Jozef Balint

President, Central Statistical Office, Hungary

It is not my intention to speak in details on the subject of this panel discussion, I only wish to add a few ideas to the analysis of the technical developments which are expected to affect statistical policy and organization. In connection with all that I am going to outline, it is to be taken into account that our thoughts and ideas were born in a socialist country where the implementation of overall social objectives enjoys first priority and technical development should be evaluated as a dependent of it, as a means.

As our view on technical developments becomes the more general, the longer period is taken under review. The contents of the forecasts for the usual 20-25 years' periods may be, in our opinion, interpreted realistically for systems of national or social dimensions, whereas for a subsystem as statistics, the effect of the technical development cannot be drawn up but with a high percentage of error even for a period of 10-15 years. Yet, we have to attempt to outline at least the expectable development.

We wish to restrict our ideas to the category of the medium-developed countries as we deem that the wider the scope of the examinations, the more general definitions can be drawn.

The concept of "medium developed countries" is not intended to be used completely in accordance with the well-known concept of international comparisons and analyses. By this concept those countries are meant, due to their level of development, to react sensibly to technical development. This means, however, to a greater extent the takeover of technical results rather than a development of unaided innovations, etc.

Consequently, the medium-developed countries are those which follow the technically most-developed countries by a time-lag of 6-8 years both in the field of introducing new development results and of disseminating them. We think that Hungary is right in considering itself one of the representatives of this category. This group of countries is expected to include several new members in the future. At this level of development, after the creation of the basis and starting institutional and operational conditions, a development policy can be elaborated in a conscious and planned way by selective use, respectively, by grading of the possibilities offered by technical development. The development policy created this way takes into account the efficiency requirements, too.

From the viewpoint of its effect on statistical organizations, we do not wish to review all the aspects of technical development, not even in an overall way. Our only purpose is to speak about one major trend which is generally characterized as "automation." The concept of automation refers here to the methodology of up-to-date decision making and its technical background. This is the field the development of which mostly determines the scientific--technical revolution in the second half of the 20th century At the same time this is the area which exercises an affect on public administration and within it on statistics.

The expected development in the field of automation has double effect in two directions strengthening each other: (1) highly efficient building elements structures and equipments are available at reasonable prices; (2) more and more complex systems are developed from them.

The development of technology and construction has entered a new stage, new results are achieved daily. The equipments operating independently and in a closed shop may reach in our days some working places; it will be, however, a general tendency in the future. Their effect will be felt with the machine tools, plotters or desks.

The mini- and micro-computers will gradually be built in the remote data terminals, magnetic tape data records etc. These equipments which were formerly independent from one another will be joined in single system.

A similar development may be experienced in the field of storage techniques Partly, the speed and storing capacity of the "traditional" storing facilities--magnetic tape, disk, drum will be considerably increased, partly, revolutionary new storing facilities will be developed.

There is a great demand for data storage in the field of statistics because for the purpose of secondary analyses all formerly collected data may be necessary in principle. For this purpose, a satisfactory solution would be comparatively cheap, but great capacity storing system which could be associated by the help of interrelation of contents. Such new techniques as laser, bubble storing, etc., which gradually leave the experimental stage, mean a promising approach to the satisfaction of such needs.

Technical devlopment in the different fields widen the evolution of computer networks. A network of this type may consist of several different computer systems joined by telecommunication lines. As a result of this trend of development, there will be an increase in flexibility mainly by the several-stage, integrated systems. In this case, the terminals of the users will more and more play the part of a small data procesing system for the satisfaction of local needs. At the same time, however, they preserve the possibility of the direct connections, if needed, with the major central systems.

These trends of development point in the direction of the planned establishment of national networks. In the reference period, this seems to be realizable for the medium developed countries, too. In an especially favourable situation are, in this respect, the socialist countries with centrally planned economies. For the establishment of such a national network and for its efficient operation, a favourable background is assured by the fact that in these countries the data supply is compulsory for each producing, servicing and social institutions. There is no obstacle to full-scope data collections.

The assumption seems to be realistic that in the future all important organization will dispose of adequate means for becoming a part of the automated statistical information system. In case of establishing a national system--

and the technical preconditions for it may be created--the local information
necessary for the preparation of central decisions will be centrally available
within a reasonable time period.

It is evident that the results of technical development as outlined above will
be felt in the medium-developed countries only to a limited degree within the
reference period. Namely, it is not only the financial means that may be
restrictive to the results of technical development but also the infrastructur-
al facilities and skills necessary to the adaptation.

The hierarachically built-up national information systems will be easily
served by a relatively smaller number of divided computer systems, too.

A computer system of this type consists of one, or more, centrally-placed
large computers which directs other medium computers operating at branch or
regional stages. At the same time they would join such servicing mini-or
micro-computers which are in relation with local and users through terminals
of different types and functions.

In the data treatment of national information systems it should be decided in
the near future which data should be part of the national system, how they are
to be gathered, who may get access to them and for what purpose.

The present autonomous and independent data base systems will shortly transform into systems of common and divided use which, due to the automation,
necessarily pave the way to the development of national information centres.

In the socialist countries, this tendency of development will be more and more
widely recognized. For the time being, in different form though, its organizational framework will be established.

Within the public administration, the statistical organizations were the first
to begin automating their data processing activity. In our days approximately
25-30 per cent of the staff of statistical organizations take a position where
they are in charge of dealing with computers and other up-to-date means.

A corporate system of this type is raising several methodological and organi-
zation-theoretical problems which had not formerly been elaborated. A national information system involves not only expedient joining of technical
instruments, but also perhaps the re-definition of the function of some administrative organs, the standardization of several activities and procedures,the
unification of concepts, classification systems used by various organs, etc.

This question has been especially important--and has been brought into the
limelight by technical development--in the common use of data bases and register systems of national importance.

Nowadays, the storing facilities of computers and the data base treating program systems render possible the establishment of data base systems/data
banks/even at the present level of development.

With regard to the increasing "hunger for information," statistical organization could--without employing automation--safisfy only a small part of
requirements.

Automation within the statistical system reacts not only upon the requirement

of information and on the connection with the data users, but also on the statistical system itself.

We examine the effect of technical progress, and particularly of automation on
the statistical system with regard to the following aspects from the point of
view of the influence on the statistical working processes and on the organization of the statistical offices.

By statistical working-process we understand the whole sphere of handling statistical data, beginning with the collection of data until putting them at the
disposal of the users.

The use of electronic computers back upon a shorter or longer past, even in
only moderately developed countries. Application of computers is, however,
different in the field of the various statistical activities. In Hungary,
too, computers were applied so far mostly for the purposes of data processing.

In recent years, however, significant progress has taken place with respect to
data storage, the handling of the different data bases by computers/organiza-
tion of data banks. The cooperation of the systems of machines for the data exchange is also solved or prepared to be solved. We have become acquainted
with microfilm technique, too,which may also play an important role in the data
storage.

Employing computers in data collection and information activity has not become
general so far. The elemination of this contradiction cannot be delayed for
long.

Technical development of data collecting may present most of the problems
Without developing the different fields of public administration, no result
might be expected here. The solution will be presented by setting up and
systematized operating of terminals, serving the primary recording and for-
warding of data. These terminals will be apt to be applied in wide fields of
economic and social activities.

The actual results in this field are indicated by the basic national
economic and social registers. These basic registers of population, territory, finances, etc. are organized for computers and made operative in a way
that they may be connected also with other information systems, so that
they may satisfy also other requirements besides the basic purpose. I should
like to mention that in the frame of the use of computers in public administration, the organization for computers of different basic registers has begun
in Hungary, too. Necessary measures have also been taken to develop the connection by computer of the individual basic registers, too. Works are proceeding which aim at the transmission of the demographic data of population
registers into the statistical information system.

It is a rather difficult task to measure and to present the effect of technical development and of the extensive application of computers on statistical
organization. Namely, this effect does not appear isolated,but interwoven
with other influences. Should we wish to imagine the connexions between technical development and the changes of the organization, the topic could probably be best approached if we spoke of the direct and indirect effects of
technical development on the organization.

A primary organizational consequence is the establishment of technical, de-

velopment, of the extending use of computers indicated in the paragraph on
statistical working process in the statistical organization, computing centre
and of a section/department/responsible for the application of data processing. Parallelly, activities of the character of computing will be accelerated also in the subject-matter units/mechanical organizators, coordinators,
etc. appear which may also have organizational consequence, particularly when
terminals connected with the computing centre are put into operation. A further effect of direct organizational character is that as a result of automatic error correction, data storage, micro-filming, etc. labour force will be
released in the professional divisions who on the one hand may be directed to
other fields, and on the other hand, persons of higher qualification/systems
analysts and designers, etc. may take their places in these units. From this
point of view, we may also speak of qualitative improvement of the given
subject-matter units.

These ideas inspire us to think about the indirect effect of technical development on statistical organization.

Quantitative and qualitative requirements for statistics are rapidly growing.
It is technical development, the extending use of computers that render it
possible to satisfy these requirements, in time, to process increasing quantity
of the data in newer and newer combinative groupings, to prepare analyses for
decision-making, etc. and to put them at the disposal of management.

As an indirect effect, the consequence of technical development appears and
can be thus valued; the presence of data processing influences the view and
approach of statisticians; using their knowledge in the field of data pro
cessing and systems analysis results in qualitative changing of professional
statistics, and thus affects statistical organization as a whole.

In countries where the statistical office has territorial bases too,--and the
socialist countries are just such--it is possible to develop division of labour in a way that part of the statistical basic activities/data collection
processing, etc. may be performed locally by means of computers. Strengthen
ing of this tendency is territorial; however, it leads to establishing computer centres organized into nationally-unified networks.

As a final result, technical development is of two-way consequences for the
statistical organization. With regard to the organization of statistics,
centralization appears as the statistical system will be unified and harmon-
ized, and the regulations centrally organized and being of general validity,
enforcement by data processing would also be achieved.

Parallel with these and at the same time, a decentralization process, too,
is taking place in statistics; its primary sign in the accomplishment of statistical basic activities is the expansion of the division of labour, broad-
ening of the division labour between the central and the placed-out parts of
statistics.

Technical and Organizational Innovations in the Informatic Statistical Field in Coming Years

C. Viterbo* and F. Marozza**

*Manager of Studies Office, Institute Centrale di Statistica
**EDP Manager, Institute Centrale de Statistica, Italy

Premise

In recent years in all countries of the world there has been an increased need of quantitative information. This need, closely linked to a faster pace in the production of goods and in the spread of education, was originally felt by mankind when it took its first steps towards the organization of community life.

So statistical information is no longer today, nor will it be tomorrow, a privileged domain of scholars and central and local public authorities. Rather, will involve ever wider levels of users, such as business, the press, individuals.

The increase in statistical demand will be accompanied by a compelling need for timeliness, reliability, and detailed analysis of information, pinpointed to areas much narrower than the traditional ones.

But a number of serious and ever-increasing difficulties now stand in the way of a speedy transfer of basic statistical information from the periphery to the center. And once the data are collected, other weighty obstacles interfere first with the data processing function, and then with the publishing of results or transmission of data to users.

All the major innovations to be introduced in coming years in the field of statistical information will have to aim at the elimination of such difficulties, both with the aid of faster and more advanced technological devices and with the adoption of organizational systems more adequate to the increased need for timeliness.

"To know in order to decide" means to have available timely statistical information before making decisions in the juridical, economic, and social fields. But in order to know, we need the cooperation of all involved and their sense of contributing something useful to themselves and to the collectivity.

We believe that the following discussion does not consist of a mere collection of risky predictions, but rather that it takes into account the hypotheses of development of data processing systems on which the entire economy is more and more dependent and conditioned.

Technical Innovations

As mentioned, progress in the statistical field in the eighties will come essentially through the work tools which technological advance has already made available or will make available, the use of which will spread with the optimization of resource management.

The use of computers in particular will grow more and more common, but in comming years it will undergo a metamorphosis involving also -- see next paragraph -- a change in organization structures.

Computers are still currently used mainly in the central stage only of the work flow. But upstream, outside this processing stage, are the data collection, control, and single information recording stages, and downstream, still off line with respect to the processor, is a statistical information distribution stage.

This system can be represented graphically as in Fig. 1.

Centralized EDP, typical of information systems of the preceding generation of processor, will certainly become obsolete in the years to come; even non-experts in information techniques have been able to note in the data processing field a trend to decentralization parallel to similar trends in other sectors of human activity, e.g., public administration.

In the field of statistical data processing, therefore, the use of numerous on-line terminals and the spreading of multiprogramming and multiprocessing will result in the reduction of the input-processing-output cycle.

Miniprocessors and terminals will be fitted into a telecommunication system to perform the work which was once centralized at the processor, i.e. the terminals, which were once low-speed teletypes. They become intelligent and will be able to operate autonomously at a local level.

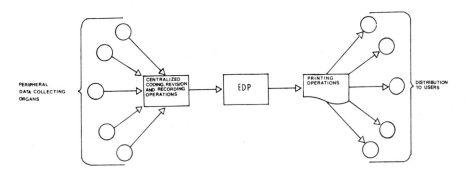

Fig. 1. Centralized EDP statistical system

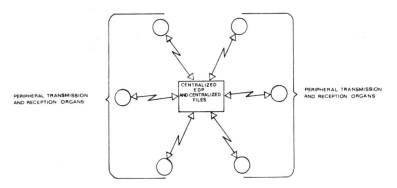

Fig. 2.   Statistical system directly connected with the
periphery

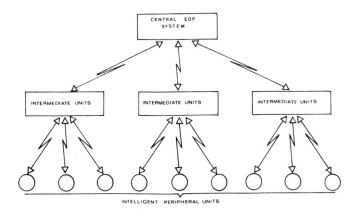

Fig. 3.   Distributed EDP system

Therefore we can foresee tommorrow's direct link-up between the periphery and the center, and vice-versa, meaning that the peripheral unit itself will perform not only the data collecting operations as in the past, but also the operations of preparing and checking the central processing units, transfering the data and obtaining the results.

It will be possible to set up all this through complex TP systems and extensive use of the peripheral units to connect the data collecting units in real time with the central processor for the transmission of information and the acquisition of results or the interrogation of centralized data bases.

The diagram is shown in Fig. 2.

It is likely that connections, on account of the high cost of lines, will not be direct ones, but rather will be organized as a system of hierarchies with intermediate concentration and collection units in clusters at perhaps regional level, as in Fig. 3.

In other words, processors will undergo a change from mere calculation centers into communication, recording, and documentation centers as well.

Such a structure will allow for the speeding up of all statistical operations, in particular those where basic information is collected by relatively small number of peripheral units (such as foreign trade, price, unemployment, transportation, and major-corporation statistics). Such a "decentralization" process has already been tested in census taking, where offset-printing techniques have also been tried out, with their attendant advantages of greater speed in printing and publishing.

Besides speed, other advantages will accompany such a connection system. For example, among these are the possibility of instant verification of the consistency of transmitted information, asking for a correction, and the processing of information as it arrives from the peripheral units so as to acquire tentative global data very quickly.

Moreover, as results flow from the central units back to the peripheral units, traditional offset techniques or more sophisticated supports can be used with the result of certainly simplifying all local data collecting and processing operations.

Statistical information should also be "distributed" between the central and the peripheral units, allowing the users themselves to carry out the local and district data collection and processing functions so as to obtain all information necessary to their peripheral activities, while the role of the central unit will be merely to coordinate and connect the peripheral units.

The individual computer will no longer fit into the picture. Rather, we shall have an interconnected system of processors exchanging data and information. Just as today's businessmen or individual savers deposit their money in the bank, so tomorrow's users will deposit their data in external files, certain of being protected and of having the possibility of remote processing.

Another sector likely to affected by technological progress is the one of the so-called data banks and automated files for data management. It could become particularly useful in specific segments of statistical activity, even if some perplexities still remain as to the productivity of the banks and files. This is especially relevant in those fields where request for data, although growing steadily, still does not cover the expenses involved in the setting up updating, and management of such banks.

The development, however, of major capacity memories with ever-decreasing unit cost per stored byte and the wider adoption of information systems and information packages will doubtless lead to a considerable enlargement of such management files in the next twenty years.

Moreover, in the future course of the transmission and diffusion of information, these will develop and be generated with exceptional intensity, creating a host of problems with their proliferation; hence, the need to think out automatic systems capable of managing them and to acquire the necessary documentation through information retrieval systems should be organized on two levels: the search for data useful to locate the required documentation and, more sophisticated, the furnishing of the documentation, if only in summary form.

Some authors have even hypothesized that the data base, as the user's fundamental asset and resource, should be designed and managed by a new individual who would have his responsibility severed from both the operational and the analytical and programming areas.

This is because such an important role should take into consideration not only the planning, designing, and setting up of the data base, but above all the adoption of standards and of safety and recovery procedures.

From the point of view of hardware, the basic trend for the coming years will be of a major development of equipment, especially for the peripheral applications and the constant search for validation and control of the input data.

Indeed, the information products for the near future -- as also a CSATA report presented by the Italian Ministry of Scientific Research points out* -- are believed to fall into the following trend lines:

-- Increased of processing speed through a wider use of miniaturization and a development of processing systems.
-- Improvement of electronic memory costs and performance through wider use of semiconductor memories and development of optical and magnetic bubble memories.
-- In the area of mass memories, a considerable development of microfilms and microfiches is forecast. They should be connected with the computers, as important supplements to the present magnetic tape of disk devices.
-- The cost of terminals will tend to decrease, and a steady proliferation of diversified and specialized terminals is foreseen,

---

*World Dynamics of the Information Systems Industry, Managemente Informatica, No5, 1976.

--In the software area, investments will reproduce the same type of development met in hardware. Such development will tend towards large-scale applications such as the following:

    1. Creation of data banks with recordings pertaining to the entire popu-
    lation
    2. Computer-assisted instruction
    3. Management of integrated information systems
    4. Economic forecasts and planning with the aid of computers
    5. Automatic large-scale documentation systems for the scientific and
    legislative sectors and for medical diagnosis.

The cost of hardware will be attributed, for the most part, to the peripheral
units. As far as the latter are concerned, this will be necessary to overcome
their performance limitations due to their mechanical nature, low speed, and
limited reliability.

The cost of software will become preponderant in comparison to hardware, with
a reversal of trend, already under way, which will continue in the years to
come.

Organizational Innovations

As already mentioned, technical and methodological innovations will also
entail considerable innovations of an organizational nature.

We can hypothesize that the future statistical information system will be
supported by a more modern and functional organization and that the bureau-
cratic structures will be improved through a speeding up of procedures and a
rational building-up of cadres.

Electronics at the service of information systems is a strategic sector under
constant development. But we can also hypothesize that the offer of advanced
technologies and the increase in the automation rate might slow down if the
problems which have sometimes prevented or decreased the use of huge resources
have not been previously solved.

Above all, there will be an evolution in the organization of the EDP statistical services. This evolution will reflect the situation of all electronic
centers; that is, those responsible for the various statistical sectors will
become responsible also for the EDP projects. They will have sole respons-
ability first for analysis and then for programming. Each statistical sector,
and in the long run the remote user himself, will have its own systems analyst
who will be responsible for the various EDP projects, which in turn will relate always to a centralized data base.

Data processing will always be submitted to centralized control; but part of
the hardware will be physically located with the user, though remote, in order
to carry out in fact the projected concept of distributed data processing.
Therefore, the users will be able, in the centralized system they helped to
generate, to directly satisfy their need for information either by the use of
generalized software or by taking direct care of their own programming, without being dependent upon the DP programming service.

This will be made easier and easier by the creation of highly advance programming languages peculiar to these applications, thus uniting in the user both the role of analyst programmer and of expert in the problems of his own statistical segment.

But in order to do that, it will be necessary to establish strict standards, especially as far as data transmission, demand control, and file management are concerned, with a careful selection of users and avoidance of duplication of files.

The role of establishing standards will essentially involve the development, creation, and maintenance of procedures. In particular, standards should relate to the following areas: development of systems, programming, operational management; documentation of activities; standards for use of equipment; format and data layout.

As we mentioned in the preceding paragraph, in increasing the potential hardware it will be necessary to proceed cautiously so as to avoid setting up processes which would later show a limited utilization rate. This can be prevented by organizing, when practicable, some form of consortium in which several users could share both resources and costs.

However, in addition to equipment and techniques, it will be necessary to deal to a greater extent with the training and motivation of personnel by laying the foundation, even from school age, for knowledge of information science, which could make the discipline accessible to a wider range of people. This would narrow the gap between demand and supply to skilled personnel, and avoid the loss of analysts, programmers, and operators from the public sector to the private sector, where better salaries are offered.

In the future the most important thing will be to avoid the drawback of processing centers having a huge computer capacity, inadequate staff, and very limited software development. In fact, software is the only thing that can keep in operation the statistical system, which, as a subsystem of the public administration system, must be integrated in the latter as a means of collection and distribution of information for the community.

Development of Informatics in the 1980's and its Influence upon the Statistical Information System*

T. Walczak

Vice President, Central Statical Office, Poland

Introduction

Informatics has a growing influence upon the trends of development and the improvement of statistical information, upon the form and organisation of statistical surveys and studies, upon the forms of rendering information to the users, on the educational composition of personnel working for statistics, as well as on the development of statistical science. Therefore, one of the major elements indispensable for working out the development program for statistical information system in the next decade has to be at least an approximate forecast of the development of informatics.

Forecasting the development of informatics does not only increase the probability of right decisions on the future form of statistical information system, but they also contribute to quicker improvement of this system. They create possibilities for earlier preparation of technical, organisational and personnel conditions for the functioning of information system.

It should be stressed at the same time that while, on one hand, the informatics influences statistical information both in its contents and its functioning, on the other hand, the development of informatics is influenced to a certain extent by requirements and needs of the statistical information development. This is an example of a system characterized by a series of interacting elements with an active feedback. This is the reason why separate forecasting for development of informatics and statistics may cause additional errors in forecasting unless their interaction is taken into account.

Anticipation of the development of informatics should be based primarily upon the observations of development of a series of interacting social and technical phenomena as well as on experts opinions.

Report devoted to anticipation of computer usage for national statistical offices in the next decade, and prepared by joint efforts of many outstanding computer specialists and statisticians can undoubtedly be treated as such a forecast.**

---

*The notion "informatics" used in this paper has been applied to denote three elements tightly related to one another: hardware, software and methods of data processing.

**Computing in National Statistical Services Beyond 1980, Computing Research Centre Bratislava, Study Group Report, July 1975.

This report, based on the analysis of the present situation, observation of the tendencies in the development of statistics and informatics is a very good starting point for many discussions on the future shape of statistical information system.

During these discussions, it should be considered particularly what achievements of informatics will have the greatest influence on statistics and what aspects of statistical information system will require improvement of changes in order to utilize at the outmost the expected development of informatics for statistical purposes. It would be purposeful to formulate statistical requirements for informatics so that new inventions and improvements in computer science could meet in greater extent the specific needs of data processing in statistics.

Development of Informatics in the 1980's

Taking into account the above mentioned considerations it seems that from the point of view of statistics the following elements of the expected development of informatics should acquire greatest importance.

In the field of computer equipment. The following basic factors will influence the improvement of computing equipment in the 80's: further technological development in production of electronic components, development in the architecture of computer system, developments of data capture technique.*

Progress in the field of electronics, in particular the technonogy of LSI circuits, miniprocessors, semiconductor memories will cause reduction of physical size of central processing units and operating memories, increase their speed, and lower the unit cost of executing basic arithmetical operations as well as the unit cost of storing the data in computer's memory.

However, at the same time computing systems are becoming more and more complex and great part of the increase in speed and price reduction is absorbed by the new functions of automatic control of these systems/complex operating systems, data transmission control/; therefore it cannot be utilized directly by the end user. The same can be applied not only to central processing units, but also to operating memories and mass storage systems. That is why, in spite of the reduction of unit cost, absolute cost of central units and operating memories remains practically unchanged.

---

*More detailed sources on the expected trends in computer hardware development can be found in numerous reports published in different countries, e.g. R.Turn: Computers in the 1980's - Trends in hardware technology, Proceedings of IFIP Congress 74, North-Holland, Pub., 1974.
In my paper attention is paid mainly to these aspects of the development of computer hardware which gave a special significance for statistics.

Improvements in the field of computer system architecture based mainly on the utilization of new electronic elements and the data transmission techniques, bring forth the tendency to create remote computing systems. These systems are characterized by the fact that irrespective of much faster and more complex central processing unit, the major part of the "intelligence" of the system is transferred to terminals, and also to the external memory control units and other modules of the whole computer system.

This development trend promotes the creation of computer networks, which are the next phase in the development of distributed intelligence system. Moreover, modern miniature electronic components create new possibilities for the hardware: it can now take over some of the functions performed earlier by software/codes conversion, some activities of the operating systems and compiler programs.

In the end of the 1980's we shall probably face the situation characterized by:

        --Common use of teleprocessing systems, utilizing very fast and complex main processing units and intelligent terminals based on mini-and micro-computers

        --Limited use of computer networks
        --Use of on-line mass storage systems/sizes in the range of 10.000 MB/which decide not only about the possibilities of storing large information files, but also about the use of modern programming systems
        --Wide use of microprocessors almost by all data processing equipment and basic office equipment
        --Specialized computer systems based on very fast reliable and small-sized minicomputers for process control, also utilized for automatization of data capture
        --New possibilities of presenting the data to the users.

In the field of software. We can anticipate further progress in the integration of the following basic elements of software: operating systems, data base management system, teleprocessing management system, selected application packages, user's languages, utility programs. Moreover, the problems of portability of software and protecting the files against non-authorized access to data will become more and more important.

As a consequence of new hardware and software facilities and the developments, new and improved "user languages" will appear. The use of the former will bring closer the end users of statistics to the vast resources of statistical information. It seems, however, that programming languages of a COBOL type will be still broadly used. It should be also remembered that, simultaneously with the improvement of programming tools which make the direct use of computers easier for the persons not qualified in the field of software, the information systems will become more complex, mainly in connection with the growing role of common data bases and the development of economic-statistical analysis methods. This will probably be the cause that the requirement for specialized personnel of designers and programmers will not be diminished, but enlarged. Only the character of work and requirements for qualifications will be liable to change.

Influence of the Expected Developments of Informatics on Statistics

Taking into account the expected trends in the development of informatics,
there arises a question of how statistics can and should utilize the achievements of the computer industry and associated improvements in the software
techniques and organisation of data processing. It seems that from the point
of view of hardware the requirements of statisticians are similar to those
of other users who apply computers to data processing.

Data processing activities in statistics differ from data processing in other
fields in scale rather than in the character of the operations performed.
This refers mainly to the requirements for central processing units, mass
storages and the basic input-output equipment. Specific requirements of
statistics are caused by the fact that performing statistical jobs we have to
deal with large volumes of input data, complex mathematical operations connected with a particularly large number of logic operations of sorting and
grouping of data and at last with a large volume of output data. At the same
time, the requirements of the end users are constantly growing in respect to
readability and form of statistical information. It can be stated, however,
that general purpose computers can fulfill most of the requirements.

Most differences between statistical and general applications exist in the
data entry field. In statistics we deal with more differentiated and complex
input documents/census forms, reports, which make the automatization of the
data preparation process more difficult.

For this reason the technical equipment for data preparation in statistics
must be more differentiated. For inputting the data from homogeneous mass
documents it will be necessary to use optical page readers capable of reading
various typewriter fonts and hand written symbols, digits and even letters.
Visual displayunits which give pictures on non-recognised characters on the
screen and which enable correction of these characters without stopping the
reading process should constitute indispensable aids.

Another basic technique of data preparation will probably be the multiple key-
board data recording systems, which will enable direct transfer of data from
the operator's key-board to magnetic media with simultaneous, automatic preliminary checking of data.

Improvement of the two basic data preparation techniques for statistics should
be expressed in gradual integration of these techniques, for instance by
equipping optical page readers with keyboards or by connecting page readers
on-line to the same computer which controls the work of multiple keyboard data
recording systems.

It should be expected that paper input data media-cards and punched tapes will
be gradually replaced by other, more perfect, mainly by magnetic media and
documents adapted to optical reading, although it is probable that during the
period discussed they shall be still in use due to a great number of punched
card equipment possessed at present by statistical offices.

The most fundamental change in the organisation of processing and presenting
of statistical data would be to resign of the presently prevailing batch processing systems, and to introduce on-line processing, where the data could be
entered to a computer directly from the places where they are produced or

collected, and the results currently presented in a direct way to the end users.

The introduction of such a system, which is similar in character to the real-time system, would mean a fundamental advancement of the whole system of storing and rendering of information. It would influence the functioning and the operative character of the system, and it would also strenghten the role of statistics in management. It would also mean the application of radical changes in the whole organisation of statistical studies and research.

Does the expected development of technology in the 80's allow us to believe that the introduction of direct processing system on a large scale would be possible?

Most probably there will be technical conditions for creating such a system. Mass storages with direct access and very large capacity of some dozen billion characters, highly developed systems of remote access to computers and advanced software will make it possible to obtain multi-aspect output results on demand.

Nevertheless, the conclusions in this matter should be formulated very carefully.

The information resources of statistical offices are already immense and there is no indication that they might be diminished in the future. On the contrary, along with the socio-economic development of the countries and the improvement of management methods the requirements in respect to statistical information will be growing.

For example, the Central Statistical Office of Poland has already about 50 billion characters recorded on magnetic tapes and disks. In bigger countries, with fully computerized statistical activities, the information resources are larger. Therefore, even gigantic computer systems equipped with the largest possible memories will turn insufficient in large countries when introduction of on-line processing into practice is considered. For smaller countries, such systems will be not available in practice because of high costs.

Moreover, it seems that in the period forecasted the development and introduction of more advanced information systems to statistics shall be more limited due to imperfection of the software, rather than due to the imperfection of the hardware.

Standard computer software designed mainly by manufacturers of computer equipment, takes into account in particular the needs of mass users of this equipment, that is the needs of industrial enterprises, financial institutions, universities, aircraft agencies, etc.

The manufacturers of computer equipment meet the needs of statistical offices in less degree, probably because these offices do not create a sufficient market for them. But in the field of software, the statistical needs are very specific when compared to other fields of computer use. That means there are more manifold interdependencies between the information comprised in data bases, there is the need to obtain variety of results which often have a complex structure, and the necessity to apply big and often very complex programs for automatic control and data editing.

This situation should cause more efforts to improve and develop its own, specific software systems for statistics. It seems that in this field, there exist particularly great perspectives for the cooperation between specialists in statistical offices in various countries.

The aim is not only to exchange experiences in software, but also to undertake joint projects on certain elements of software, especially when creating and using statistical data banks is considered.

Another task for statisticians is to improve the language used to describe statistical information resources, to unify economic categories applied to statistics and planning, and to precise definitions and concepts used in statistics.

These works constitute one of the elements of the methodological improvement of surveys and statistical studies which ensure internal compactness of statistical information system. They contribute much to the elaboration of more developed software systems of statistical data banks. They play also an important role in ensuring methodological consistence of statistical information with other information systems in the country.

The improvement of computer hardware and software will create possibilities for the introduction of fundamental changes in the organisation of data processing in statistics. These changes will be much influenced by changes in technology of data collection in the basic units outside statistical organs, that is in industrial enterprises, administration, educational establishments, financial institutions, medical services, population registers, etc.

There will be conditions for the introduction of changes to the methods of supplying the data to statistical offices in order to increase the role of obtaining the data directly from computerized systems of enterprises. Some data, which are now received by statistical offices in the form of special reports prepared by enterprises, will be obtained in the form of machine readable media, being by-products of data processing of the basic evidence.

The introduction on a wider scale of direct flow of information from enterprises to statistical offices in the form of machine readable media will make it possible to lessen manual work both in units which supply data, and in statistical offices. It will give a chance to obtain more detailed data and more precise source information. However, at the same time the introduction of such a direct feeding system will require great effort to organize a new system of information inflow, to ensure methodological consistence of the information system of enterprises with statistical system, to create adequate law regulations which define the responsibility for the timeliness and reliability of information, to overcome the difficulties with the ability to transfer the information form different media.

Most probably, the above-mentioned difficulties will cause that the reporting by economic units will be the leading general form of obtaining data for statistics in the 80's.

Broad use of teleprocessing in forecasting period and the parallel improvement of software which ensures the use of computers by statisticians who have no special training in computer science can change in an essential way the character of a job of a statistician. He will be less dependent on a computer specialist, because he will be able to receive all the indispensable data for the current analysis and calculations. The barrier between statisticians and computerized information resources existing at present will be partially liquidated. This will create much better conditions for the statisticians for deeper analysis of socio-economic phenomena, but on the conditions that they notice these possibilities in time and that they will be prepared for their utilization.

G

Technical Developments expected to Affect Statistical Policy during the 1980s

I. Ohlsson

Director General, National Central Bureau of Statistics, Sweden

Introduction

Technical development in the field of statistics is conditioned by new demands
for information, as well as by new premises created by the development in
other fields, calling for new statistical methods etc. Techniques and methods
adjusted to the special requirements of statistics will probably, to a great
extent, have to be developed by the national statistical offices themselves
It is likely than an international exchange of experience will as previously
play a decisive part.

Legal and other institutional restrictions provide the framework within which
the techniques are developed. To some extent, such frameworks per se imply
demands for development of techniques and methods. Methods to ensure confi-
dentaility and secrecy in connection with storage and processing of computer
registered material can be mentioned as an example of regulations imposed on
the statistics producer by the community.

The exposition in this paper of the expected technical development during the
1980's is largely based on the experience and trends of today's statistics production in Sweden. Possible changes in the institutional conditions as well
as technical innovations etc. naturally constitute elements of uncertainty
in the presentation. Furthermore, the potentiality of the techniques and
methods treated in this paper is to some extent difficult to estimate.

"Technical development" is here interpreted fairly widely. EDP-technology
(both hardware and software), statistical methods (including measurement
methods) and techniques to coordinate statistical materials and to utilize
administrative materials for statistical purposes are all included. By "statistical policy" is here meant both decisions as to which services a national
statistical office is to make available (i.e. the contents and volume of the
statistics) and the manner in which statistical information is to be collected,
processed, analyzed and presented.

Development of production systems

Since the middle of the 1960's, the National Central Bureau of Statistics (SCB)
in Sweden has devoted systematic efforts to developing the statistical production systems. In comparison with the traditional production systems, the object is to achieve systems characterized by:

--Increased capability to make statistical processings adjusted to the actual demands for statistics

--A better comprehensive view of the total statistics production for both users and producers

--Increased flexibility which permits combinations of data from various sources

--Improved and speedier access to data stored in the statistical data bases (inter alia statistical time series)

--Increased use in the production of statistics of data primarily collected for administrative purposes.

To reach these objectives, it is necessary to coordinate definitions of variables and units as far as possible, and to draw up common identification systems. Techniques for easily accessible documentation of the data stock have to be developed as well. The processing technique has to be developed in respect to data base handlers and standard elements for certain functions in the statistical production systems. Some such functions are e.g. checking, editing and coding, as well as table specification and production.

A better adjustment of the production systems to the users' demands requires techniques to describe the statistics produced, techniques to enable the users to specify the statistical outputs and techniques to distribute the statistics to the users in desired manners.

Development of Standards and Other Coordination Instruments

Data registered in administrative systems (data bases) are expected to become increasingly utilized for statistical purposes. For coordination between statistical and administrative data bases (as well as between various statistical data bases) uniform identification systems and integration keys are required. The confidentiality requirements do impose restrictions in the identification of single statistical units and the linking of statistical materials. Such conditions might probably be partly satisfied through suitably designed security systems etc. For the coordination of statistics there is, among other matters, a need to develop simple code systems for the primary registration of variables and units. In some cases, "data transformers" to transform information in administrative data bases to a form suitable for statistical processes might have to be developed.

Production coordination also requires basic instruments in the form of e.g. catalogues of variables, basic nomenclatures and classification standards in several subject-matter fields. By "catalogue of variables" is meant a homogeneous and easily accessible documentation of the definition of statistical units, variables etc. in existing data bases.

Today, many studies are based on sample surveys. In studies covering the same population (of e.g. enterprises), it is possible to design the samples in such a way that either the same units are surveyed in two or more studies (positive coordination) or as few units as possible are common to the studies (negative coordination). Such methods for sample coordination have to be introduced and have perhaps also to be further refined.

Presentations of the quality of statistics are required to improve basic decision materials and lessen the risks for erroneous decisions. To the extent
that the statistics users themselves are to design their outputs at computer
terminals, the formulation of the quality statements poses special problems.

It can be generally assumed that the cost of hardware will not remain the
limiting element it has previously been. Rather it is the development costs
that to a great extent will limit the possibilities to change and develop production systems. This means that increased use of standard elements (standard programs and standard systems, e.g. data base handlers) becomes desirable.

The possibilities to give the statistics users easy access to statistics
as well as the possibilities to speed up the editing procedures in the statistics production largely depend on the price/performance development for
random-access stores. The development seems to move fairly rapidly, which
means that considerably more data than before might be kept available on-line.

The tele-communication technique is expected to be further developed. This
means increased possibilities to transmit automatically statistical material
in computer-readable form from the statistics producing systems to the users'
computer systems for planning or research. Similarly, it will become possible
to get some input data to the statistics directly via the tele-communication
network, inter alia from administrative systems.

The development of data base handlers is expected to continue and accelerate.
Standardizing work might perhaps lead to a module approach in the future, so
that modules for various functions can be combined. For instance, it might be
desirable to combine the terminal handling function from a commercially available data base handler with own modules for data description etc.

In data base handling systems, display terminals are expected to be more and
more extensively used for faster and more effecient (integrated) checking,
editing, coding and other manual processing of statistical primary materials.
Computer-supported manual processing of primary data is in general assumed to
become increasingly common. Some suppliers of data might also come to have
direct contacts with the statistics producer via data terminals.

Today, systems are already being developed for on-line service to some statistics users who demand fast and flexible output data. By means of these systems the users are offered the facility of output of tables etc. directly via
terminals (preferably equipped with display consoles). Interactive techniques
for communication with the computer is assumed to be greatly used. All factors considered, this trend will probably be intensified during the 1980s.

The rate of introduction of new technology for the input of data to computer
systems has been such that it is very difficult to choose the best technology
and methodology even for a particular set of data, let alone the optimal technological investment for the input of all the sets of data of a statistical
office.

Continued improvements in the techniques already available are expected - key-
punch, key-edit, OCR, direct input via computer terminals with its possibilities for on-line editing and coding etc. However, the possibilities for entirely new techniques seem limited. Instead, methodological studies to ensure
the best use of the various techniques have to be undertaken.

By the 1980s, it is hoped that experience has given a full understanding of the costs and errors involved in each technique.

Among the costs of any technique are its effects on the people who use it, and there will probably be a tendency away from the dehumanizing type of work inherent in key-punch or key-edit input towards the more involved and pleasant work of conversational data entry, using computer terminals and interactive editing programs.

Methods for automatic (or semi-automatic) coding are being developed and introduced in the statistics production. As indicated previously, automatic coding can be carried out in direct association to the necessary manual handling of the primary data. The simple, routine coding is then carried out by the computer, while more complicated cases are handled by the operator.

At present the checking of primary data is mainly carried out manually and/or automatically. The correction of erroneous data on the other hand is almost exclusively non-automatic. However, a few statistical processing systems include custom-made automatic sub-systems for checking as well as supplementation and editing through imputations of data which are missing or have been rejected by the editing program. The development of standardized, entirely automatic checking and editing systems might become intensified.

Certain well-known estimation methods to utilize supplementary information are very well suited for a database-oriented processing environment. These estimation methods have to be implemented and perhaps also further developed.

The on-line service previously mentioned presupposes the development of techniques to protect statistical primary data against unauthorized access, inter alia authorization systems. The development of such systems is closely associated with the development of operating systems. Methods for encryption, file-splitting etc. have to be developed as well. Techniques to prevent disclosures in tables etc. have to be developed and implemented in systems for output of statistics via terminals. For instance, the techniques for suppression or intentional distortion of data so far tested have to be improved and further developed.

Today, there already exist a great number of standard programs and software packages for tabulation, multivariate statistical analysis and graphic presentation of statistical data. This kind of software will probably be used to a considerably greater extent than has sofar been the case to process statistical material. It is conceivable that the programs might be put to the statistics user's direct disposal at the terminal (which perhaps will be equipped with a minicomputer for the necessary computations).

Display terminals, on-line services and services referred to in the previous paragraph, presuppose that software for man-computer communication is developed. This is applicable on e.g. interactive programming as well. Some progress has been made and substantial development in this field can be expected.

Develpment of Measurement Methods

As indicated above, the statistics producer will probably to an increasing ex-

tent use data from administrative systems for traditional fields of statistics.
For instance, population and housing censuses might largely be given the form
of "register censuses". However, a development of new statistics fields is
expected too, requiring separate collections of statistical primary data. Two
such fields, where statistical information is at present scarce, are health
and working environment and natural environment. For these and other statistics fields, measurement methods have to be developed. The same is the case
as regards measurements of attitudes and knowledge required for social, cultural and economic planning.

Methods also have to be developed to master measurement and non-response problem in the future production of statistics. Particularly, the non-response
problems tend to become more and more difficult to manage in surveys of individuals and households. Data collection methods and statistical methods
have to be developed in order to keep the non-response rate at an acceptably
low rate and/or to reduce the effects of inevitable non-response. Some examples are on one hand techniques to satisfy the integrity and confidentiality
requirements of the respondents, on the other methods for estimates based on
combinations of data from several materials originating from the same or different respondents.

Other developments

The ultimate effect of current work in the foundations of sampling theory and
statistical inference on the practice of official statistical offices is difficult to predict. It does seem safe to say, however, that as statistical
offices learn more about their own measurement and estimation procedures--
particularly in the context of more integrated systems-- they will become
bolder and more imaginative in the use of this information for the improvement
of both the quality and the timeliness of current estimates. In particular,
Bayesian methods might be more used, with prior distributions supplied from
the stock of data on the populations under study and on the errors, in the
measurement procedures, for the production of official statistics.

Developments can also be expected in other methods of importance for statistical policy but not previously mentioned. For instance methods of surveys
and analyses of user demands, cost/benefit analyses for evaluation of the
public benefit of statistics, and models for the total statistical production
process in society (also to the extent it is not directly conditioned by e.g.
the technological development).

A BLS Approach to Designing Computer Systems to Compile Statistics

R. C. Mendelssohn

Assistant Commissioner, Bureau of Labor Statistics, United States Department of Labor

The Bureau of Labor Statistics (BLS) started to switch to large-scale, third-generation computers six years ago. Our deliberate approach has given us time to study our past mistakes and size up new opportunities. We have taken the painful step of going from one machine generation to another before. In those cases, some 20 to 40 major systems (depending on how you count) were moved, one by one, in a patchwork fashion. The result did not serve a Bureauwide end and, in fact, they often even failed to meet the needs of each separate system.

This time we have worked out a blueprint, a master plan, to give unity to the BLS approach to third-generation computers. The work is partly done. Some will take a great deal more time and some may never be finished.

The Plan: Data Base Management   The Central Ring

Important to understanding, the BLS picture is the magnitude of its data base. The product of our surveys is mostly time series. There are nearly 30,000 in the Bureau's data base and they include important measures of the Nation's economic well being, such as the Consumer Price Index, the Wholesale Price Index, and the unemployment figures. Cross-sectional summary data are also stored in the data base. These include occupational wage information, economic projections, and productivity studies, for example. In addition, there are all the reports we get from industrial and commercial establishments, which we use to compile these summary statistics. We get over a quarter-of-a-million of these micro reports each month.

The plan we worked out has as its central notion a data base management system to process the Bureau's macro and micro data elements (Fig. 1). We bought the data base manager called TOTAL, rather than working out our own. This software allows us the use of network concepts to set down logical relationships, somewhat like the way a card catalog points to the shelf location of books in a library. A search for data follows a path to the immediate access disk hardware where the item can be found, rather than following a sequential search from the beginning of the magnetic tape file. As a result, we can process reports promptly, and need not hold them for a batch. Although mainly keyed to production requirements, the management system also allows our researchers to get directly to the data they want.

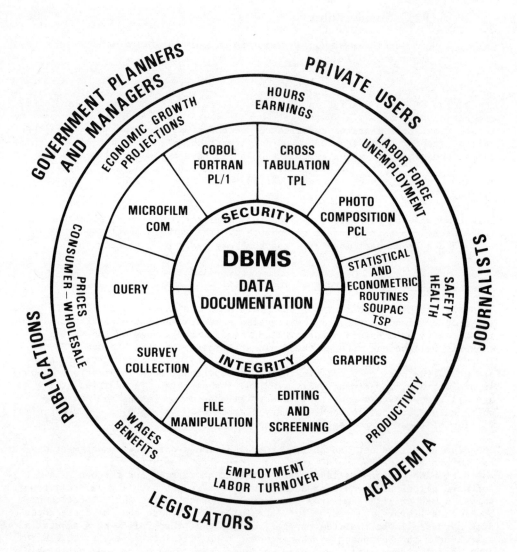

Fig. 1

Data Base Codebook and Dictionary: Data Documentation

You and I and computers have common problems when we seek to process data: We
all need to know what is available, which data are to be processed, where they
are located, and how to get them. Whether we are a computer programmer or a
research analyst, we must identify the variables we intend to use. No matter
which programming language we use, we must somehow map a path to the right
places in various storage media to locate the data that our programs must ex-
tract and process.

The BLS goal is to provide a "Codebook" and a "Dictionary," which we and the
machine can read as sources of most of this information. For our researchers,
there will be a Dictionary which names what is in the Data Base and defines
the variables cited. For the programmer, there is a Codebook with technical
data needed for describing the data formats and locations. Together, the
Codebook and Dictionary from a "Data Documentation" system. A version of the
Codebook is now in use. Work on the Dictionary will probably begin this year.

The Second Ring: Security and Integrity

Responses to Bureau questionnaires are made under a pledge on our part to keep
these data in confidence, to be used only for statistical purposes. We take
pains to insure that our pledge is kept. Therefore, we must surround the BLS
Data Base with a circle of protective software.

The protective software must recognize three types of accessibility to the
core: The first is to data that are published. These are available to anyone
in BLS. In the long run, we would hope to make these data accessible to anyone, whether inside or outside of the Bureau, in or out of the Government.
The second need is for access by BLS employees to data that are not published.
They include summary figures that contribute to published information but are
not statistically reliable on their own account. Usually, these figures are
accessible only to the professional staff that is responsible for the figures.
Well inside the data base, in a third level of accessibility, we find the respondent reports withheld and protected to keep our pledge of confidentiality.

Some of the software to carry out these restrictions and protect against unauthorized access is in place. Besides guarding against unauthorized use,
software surrounding the data base must also protect the integrity of the figures. By this, I mean we must be sure that we can recover from equipment fail-
ures, such as a head crash on a magnetic disc or inadvertent erasure of a magnetic tape. Work in this area relies largely on software provided by the computer center used by the Bureau.

The Third Ring: Modular, Gereral-Purpose Statistical Processing Programs

In the plan to use third-generation equipment, some of the 50 or 60 "systems"
were combined because similarities could be incorporated into a single approach. For example, we have had separate statistical surveys for collecting
prices of commodities in the wholesale market and in international markets.
These data contribute to our Wholesale Price Index, Industry Sector Price Indexes, and International Price Comparison programs. If we had gone the old
way, these separate surveys would have prompted separate, independent computer

systems, even though the calculations and other computer operations were often essentially the same. In the redesign effort, the three systems were reduced to one, at considerable savings in time, money, and subsequent maintenance. A similar approach was taken in combining four recurring occupational wage studies into one computer system.

The notion of combining systems with like functions brought us the idea of generalized software to deal with these functions and replace ad hoc programming. In our search, we found that common functions are indeed shared among many Bureau survey systems and that it was possible to identify these rather easily. Thus, our third ring holds general-purpose programs that are tailored to the unique needs of large-scale, statistical data processing.

Although readily identified, we also found that the construction of a full line of functional modules is a tall order. More than a dozen general-purpose programs are required. Building these modules is a costly, complex task and will take time to complete. Our approach is to take a step at a time. As resources permit, we are directing our efforts toward the modules that are most needed and that will bring prompt payback.

Now, let's take a look at the functional modules that make up the third ring.

Table Producing Language: TPL

The first outcome of the step-by-step approach was a powerful new computer language that allows Bureau staff to turn out statistical tables at less cost and more promptly than the traditional computer languages allow. It is called Table Producing Language (TPL). The TPL system already knows what a table is and how to generate one. It only needs to be told about the one wanted. Thus, when one describes the table wanted with the table producing language, one avoids the tedious and time-consuming effort otherwise involved in telling the computer, step by step, how to make the calculations and how to lay out the table. This approach has severed the connection between the user and the computer. The user need not be familiar with how the computer works. Moreover, it allows Bureau social scientists to use the everyday common BLS nomenclature to describe the tables. In short, TPL has reduced a burden, speeded work, and increased the BLS capacity to respond.

Print Control Language: PCL

One important BLS goal is to produce statistical tables for analysis, leading to reports on economic affairs, and for publication in periodicals and bulletins. In fact, statistical tables form the bulk of our printed output. An extension of TPL serves table production in ways other than as a tool for tabulation. We need the results on a clean copy for photo-offset printing. For this purpose, we must surround the numbers with a clean and concise framework of explanatory alphabetic information, such as table and column headings and stubs, as well as footnotes and similar data that make sense and are readable. We call this module the Print Control Language (PCL).

Figure 2 shows an example of PCL output. Notice the overstrike in the title. Through PCL, the user can specify column and stub widths and insert his own alphabetic labels for each variable, as well as footnotes and other text.

Table 18. Shift differential provisions

(Percent of production workers in miscellaneous plastics products establishments by shift
differential provisions, United States, selected regions and areas, September 1974)

Shift differential 1	United States 2	New England	Middle Atlantic	Border States	Southeast	Southwest	Great Lakes	Middle West	Pacific
Second shift									
Workers in establishments with second shift provisions	93.5	95.5	87.7	87.2	96.5	95.4	96.1	96.8	98.2
with shift differential	86.9	86.5	85.8	76.2	84.9	80.1	90.8	89.7	85.0
Uniform cents per hour	76.0	70.1	75.6	76.2	72.6	77.7	85.6	80.1	86.6
Under 5 cents	1.2	-	.5	6.6	-	2.3	2.0	-	-
5 cents	12.2	4.8	10.8	8.0	19.6	12.8	12.6	25.5	9.0
6 cents	2.0	1.6	.5	8.7	1.4	1.7	3.0	3.6	-
6.5 cents	.2	1.7	-	-	-	-	-	-	-
7 cents	2.6	1.9	1.8	5.0	3.2	1.4	4.1	-	-
7.5 cents	.4	-	.2	-	-	4.3	.3	-	-
8 cents	4.6	3.0	1.8	4.7	3.4	2.1	8.5	2.7	1.3
8.5 cents	.2	-	-	-	-	-	.6	-	-
9 cents	1.9	3.2	1.8	9.3	3.1	-	1.6	-	-
9.6 cents	(3)	-	-	-	-	-	(3)	-	-
10 cents	27.2	34.1	28.8	14.2	26.5	34.1	27.6	28.1	13.0
11 cents	1.4	.8	-	-	-	-	3.3	1.4	1.0
12 cents	6.2	.9	3.9	8.2	3.5	2.0	12.0	2.7	1.7
Over 12 and under 15 cents	3.7	3.5	10.7	5.9	-	-	1.7	2.3	.5
15 cents	8.0	11.7	7.2	2.2	6.9	11.4	6.9	9.3	12.4
Over 15 and under 20 cents	1.1	-	3.0	-	-	-	.1	-	3.9
20 cents	1.7	1.9	2.8	3.3	1.1	3.6	.9	4.5	-
Over 20 and under 25 cents	.3	-	.8	-	-	-	-	-	1.3
25 cents	.7	-	.9	-	3.9	2.1	-	-	-
Over 25 cents	.6	1.1	.3	-	-	-	.6	-	2.5
Uniform percentage	7.6	16.4	10.2	-	10.3	-	4.4	5.2	11.1
Under 5 percent	.5	-	-	-	3.3	-	.6	-	-
5 percent	2.5	10.0	2.5	-	.8	-	2.0	-	1.8
Over 5 and under 10 percent	.7	1.6	-	-	-	-	.5	2.6	2.3
10 percent	3.5	4.8	6.7	-	6.3	-	1.2	2.6	4.3
Over 10 and under 15 percent	.1	-	-	-	-	-	-	-	1.0
15 percent	.4	-	1.0	-	-	-	-	-	1.7
Other	3.2	-	-	-	2.0	2.3	.5	4.4	27.3
Third or other late shift									
Workers in establishments with third or other late shift provisions	87.7	90.9	83.9	87.2	84.4	83.9	91.9	90.6	82.6
with shift differential	83.2	86.0	83.1	76.2	74.4	74.3	87.1	84.4	81.8
Uniform cents per hour	73.1	70.2	73.1	76.2	62.8	72.0	83.2	77.4	45.0
5 cents	.8	.7	.7	-	3.1	3.3	.3	-	.6
Over 5 and under 8 cents	1.7	.9	-	-	1.2	1.9	4.0	-	-
8 cents	1.6	-	..	6.6	2.1	5.1	1.3	5.7	-
9 cents	.3	-	-	-	-	-	.4	3.6	-
9.5 cents	.1	1.6	-	-	-	-	-	-	-
10 cents	15.0	10.2	13.6	14.2	17.6	11.5	18.7	17.8	8.6
11 cents	.4	-	-	2.0	2.1	-	.5	-	-
12 cents	3.2	-	1.7	5.0	1.7	3.8	6.2	1.1	-
Over 12 and under 15 cents	3.5	3.1	3.6	12.0	.6	-	4.5	4.2	1.1
15 cents	17.6	14.2	17.8	14.2	15.5	15.8	19.9	26.7	9.9
16 cents	3.3	3.1	1.9	4.4	3.5	2.0	5.2	2.7	-
17 cents	2.0	-	.4	6.0	3.1	-	3.8	-	1.0
18 cents	1.3	1.0	1.5	1.7	-	-	1.7	2.3	.7
19 cents	1.0	1.7	1.3	-	-	-	1.0	2.7	1.0
20 cents	11.9	21.2	15.3	-	8.6	16.3	10.2	6.2	8.4
Over 20 and under 25 cents	1.6	-	5.2	-	-	-	.9	-	1.2
25 cents	3.8	3.5	5.6	3.3	3.5	7.7	2.9	-	4.1
Over 25 and under 30 cents	.8	3.8	.5	5.9	-	-	-	-	1.1
30 cents	1.5	4.0	.9	-	-	4.6	1.4	-	2.9
Over 30 cents	1.5	1.1	2.3	-	.5	-	.5	4.5	4.3
Uniform percentage	6.3	15.8	9.0	-	9.5	-	3.4	2.6	7.0
5 percent	.3	-	.4	-	-	-	.5	-	-
Over 5 and under 10 percent	1.1	1.6	.3	-	3.3	-	.9	2.6	1.2
10 percent	3.0	12.7	3.9	-	4.4	-	1.4	-	.8
Over 10 and under 15 percent	.3	.6	.9	-	-	-	-	-	-
15 percent	1.1	-	2.4	-	1.8	-	.6	-	2.2
Over 15 percent	.6	.9	1.0	-	-	-	-	-	2.8
Other	3.7	-	1.1	-	2.0	2.3	.5	4.4	29.8

1/ Refers to policies of establishments currently operating late shifts or having provisions covering late shifts
2/ Includes data for the Fountain region in addition to those shown separately
3/ Less than 0.05 percent

NOTE: Because of rounding, sums of individual items may not equal totals.

Fig. 2.

Decimal points and special symbols, such as dollar signs, are available.
There are options for hyphenating and centering titles. Taken together,
these features can create tables that are acceptable for direct photo-offset
printing.

Even though we can derive tables that are acceptable for printing, we have
extended the print facility one more step. Reproduction of computer printout
is less satisfactory than output from an electronic photocomposition device.
Heretofore, such devices required a new tailor-made computer program for each
table to be photo composed. A new version of the Print Control Language
permits tables to be formed by a photocomposer (Figure 3) without the need to
write computer program. The user has a wide range of print size, style, and
other choices for typographic enhancement with the result that printed tables
appear to have been typeset.

Statistical and Econometric Routines: SOUPAC, TSP

The basic TPL arithmetic calculations do not permit many of the complex, scientific analyses required by the Bureau's statisticians, economists, demograph-
ers, and other analysts. There is an option that allows the research user
to shunt tabulated results into a package of statistical routines called
SOUPAC that offers analysts a broad range of statistical procedures, including the most common, and still allows BLS researchers to undertake complex
statistical tasks.

In a similar way, it is intended that other packages will be provided to
accept and process data from the data base. For example, the Bureau has
acquired a copy of the Harvard-MIT version of the econometric package called
Time Series Processor (TSP). It is designed to carry out all the computational steps which occur routinely in statistical analysis of time series and also
to provide some of the more sophisticated econometric techniques for manipulation and analysis of time series.

General-Purpose Charting System

The Bureau has acquired a generalized time-series charting system. Use of
the system as designed requires some programming knowledge. To simplify
things for our users, it will be embedded in a very high-level language in the
same way that cross-tabulation routines are at the core of the Bureau's Table
Producing Language. The enhanced system will allow users to produce time-
series charts, editorially suitable for publication. A wide choice of features are planned to permit the user to have control over the appearance of
the chart so that it can be photographed for photo-offset printing without
further editing.

Screening, Editing, and Correction System

BLS systems analysts and programmers specify and write many computer programs
for testing the accuracy of data reported to us by our survey respondents and
mark those which appear to be of doubtful validity or clearly erroneous. Although the subject matter may differ sharply from survey to survey, the computer steps and factors taken into account are essentially the same for all
surveys. By combining the requirements for reviewing statistical data into

Table 18. Shift differential provisions

(Percent of production workers in miscellaneous plastics products establishments by shift differential provisions, United States, selected regions, and areas, September 1974)

Shift differential[1]	United States[2]	New England	Middle Atlantic	Border States	South-east	South-west	Great Lakes	Middle West	Pacific
Second shift									
Workers in establishments with second shift provisions	93.5	95.5	87.7	87.2	96.5	95.4	96.1	96.8	94.2
With shift differential	86.9	86.5	85.8	76.2	84.9	80.1	90.4	89.7	85.0
Uniform cents per hour	76.0	70.1	75.6	76.2	72.6	77.7	85.6	80.1	46.6
Under 5 cents	1.2	-	.5	6.6	-	2.3	2.0	-	-
5 cents	12.2	4.8	10.8	8.0	19.6	12.8	12.6	25.5	9.0
6 cents	2.0	1.6	.5	8.7	1.4	1.7	3.0	3.6	-
6.5 cents	.2	1.7	-	-	-	-	-	-	-
7 cents	2.6	1.9	1.8	5.0	3.2	1.4	4.1	-	-
7.5 cents	.4	-	.2	-	-	4.3	.3	-	-
8 cents	4.6	3.0	1.8	4.7	3.4	2.1	8.5	2.7	1.3
8.5 cents	.2	-	-	-	-	-	.6	-	-
9 cents	1.9	3.2	1.8	9.3	3.1	-	1.6	-	-
9.6 cents	(3)	-	-	-	-	-	(3)	-	-
10 cents	27.2	34.1	28.8	14.2	26.5	34.1	27.6	28.1	13.0
11 cents	1.4	.8	-	-	-	-	3.3	1.4	1.0
12 cents	6.2	.9	3.9	8.2	3.5	2.0	12.0	2.7	1.7
Over 12 and under 15 cents	3.7	3.5	10.7	5.9	-	-	1.7	2.3	.5
15 cents	8.0	11.7	7.2	2.2	6.9	11.4	6.9	9.3	12.4
Over 15 and under 20 cents	1.1	-	3.0	-	-	-	.1	-	3.9
20 cents	1.7	1.9	2.8	3.3	1.1	3.6	.9	4.5	-
Over 20 and under 25 cents	.3	-	.8	-	-	-	-	-	1.3
25 cents	.7	-	.9	-	3.9	2.1	-	-	-
Over 25 cents	.6	1.1	.3	-	-	-	.6	-	2.5
Uniform percentage	7.6	16.4	10.2	-	10.3	-	4.4	5.2	11.1
Under 5 percent	.5	-	-	-	3.3	-	.6	-	-
5 percent	2.5	10.0	2.5	-	.8	-	2.0	-	1.8
Over 5 and under 10 percent	.7	1.6	-	-	-	-	.5	2.6	2.3
10 percent	3.5	4.8	6.7	-	6.3	-	1.2	2.6	4.3
Over 10 and under 15 percent	.1	-	-	-	-	-	-	-	1.0
15 percent	.4	-	1.0	-	-	-	-	-	1.7
Other	3.2	-	-	-	2.0	2.3	.5	4.4	27.3
Third or other late shift									
Workers in establishments with third or other late shift provisions	87.7	90.9	83.9	87.2	84.4	83.9	91.9	90.6	82.6
With shift differential	83.2	86.0	83.1	76.2	74.4	74.3	87.1	84.4	81.8
Uniform cents per hour	73.1	70.2	73.1	76.2	62.8	72.0	83.2	77.4	45.0
5 cents	.8	.7	.7	-	3.1	3.3	.3	-	.6
Over 5 and under 8 cents	1.7	.9	-	-	1.2	1.9	4.0	-	-
8 cents	1.6	-	.6	6.6	2.1	5.1	1.3	5.7	-
9 cents	.3	-	-	-	-	-	.4	3.6	-
9.5 cents	.1	1.6	-	-	-	-	-	-	-
10 cents	15.0	10.2	13.6	14.2	17.6	11.5	18.7	17.8	8.6
11 cents	.4	-	-	2.0	2.1	-	.5	-	-
12 cents	3.2	-	1.7	5.0	1.7	3.8	6.2	1.1	-
Over 12 and under 15 cents	3.5	3.1	3.6	12.8	.6	-	4.5	4.2	1.1
15 cents	17.6	14.2	17.8	14.2	15.5	15.8	19.9	26.7	9.9
16 cents	3.3	3.1	1.9	4.4	3.5	2.0	5.2	2.7	-
17 cents	2.0	-	.4	6.0	3.1	-	3.8	-	1.0
18 cents	1.3	1.0	1.5	1.7	-	-	1.7	2.3	.7
19 cents	1.0	1.7	1.3	-	-	-	1.0	2.7	1.0
20 cents	11.9	21.2	15.3	-	8.6	16.3	10.2	6.2	8.4
Over 20 and under 25 cents	1.6	-	5.2	-	-	-	.9	-	1.2
25 cents	3.8	3.5	5.6	3.3	3.5	7.7	2.9	-	4.1
Over 25 and under 30 cents	.8	3.8	.5	5.9	-	-	-	-	1.1
30 cents	1.5	4.0	.9	-	-	4.6	1.4	-	2.9
Over 30 cents	1.5	1.1	2.3	-	.5	-	.5	4.5	4.3
Uniform percentage	6.3	15.8	9.0	-	9.5	-	3.4	2.6	7.0
5 percent	.3	-	.4	-	-	-	.5	-	-
Over 5 and under 10 percent	1.1	1.6	.3	-	3.3	-	.9	2.6	1.2
10 percent	3.0	12.7	3.9	-	4.4	-	1.4	-	.8
Over 10 and under 15 percent	.3	.6	.9	-	-	-	-	-	-
15 percent	1.1	-	2.4	-	1.8	-	.6	-	2.2
Over 15 percent	.6	.9	1.0	-	-	-	-	-	2.8
Other	3.7	-	1.1	-	2.0	2.3	.5	4.4	29.8

[1] Refers to policies of establishments currently operating late shifts or having provisions covering late shifts.
[2] Includes data for the Mountain region in addition to those shown separately.
[3] less than 0.05 percent.

NOTE: Because of rounding, sums of individual items may not equal totals

Fig. 3

one general-purpose computer program, the need to continually write special-
purpose, tailor-made programs will be reduced. Tailor-made programs demand
substantial resources and impede timeliness. A general system can make the
Bureau more responsive to public demand for speedy compilation of statistical
information.

Our success in designing and implementing a generalized cross-tabulation system, some analysis of a generalized approach to screening, as well as work by
Statistics Canada, suggests that the project is feasible.

File Manipulation System

The situation where the reported data, even when correct, are not ready for
tabulation is a common one. Typically, at the beginning of a survey, information is organized for easy and efficient collection, entry, and screening.
The resulting file organization is not usually suitable for tabulation and
subsequent analysis. The capability to reorganize files will be incorporated
into a proposed system, which we call the General-Purpose File Manipulation
System.

Its capabilities will include selecting subsets of the original data, merging
data from different files, reorganizing the data structures, and calculating
new variables. This system will fill the gap between the general screening
system, which is concerned with the correctness of reported data, and TPL,
which tabulates and displays data. Work on the file manipulation system is
scheduled to begin later this year.

General Survey Collection System--Registers and Sample Selection

To provide a general solution to processing problems that go with mailing and
control of responses to our questionnaires, a module, designated as a General
Survey Collection System, is planned. It will provide sample selection,
mailing label generation, response control, follow-up list generation, and
status reports. To serve these ends, the system must support and maintain
registers of names and addresses and other information about respondent attributes, such as industry code and employment size. A major factor in this
system is the requirement to maintain about 4 million names, addresses,
Standard Industrial Classification (SIC) codes, size, and other attributes of
establishments that comprise the Bureau's universe file from which sample
cases are selected. Work on the universe file management system is under way
and is expected to be completed in about a year.

Query

From time to time, there is interest in retrieving records supplied by a
specific Bureau respondent or sets of respondents. Such interest may derive
from discussions with the respondent about his report or from concerns about
the individual case, for example, in comparing responses with an independent
quality measurement survey. A retrieval facility to query the micro data
base would serve this need.

Microfilm (COM)

Computer output on microfilm (COM) is practiced in BLS. However, it is not
clear at this point that a general solution for placing computer output on
microfilm is an important Bureau requirement. At present, printed tables are
the principal BLS output. In the event that a generalized COM tool is justified, it will be an extension of the Print Control Language System, which has
been implemented with this possible enhancement in mind.

Application Systems: PL/1, FORTRAN, COBOL

Many of you would argue that data processing requirements of national statistical agencies cannot be totally generalized, because all needs cannot be
anticipated. I agree. We know that the compilation of some statistics is
based on special mathematical and statistical arguments that have application
only in a specific instance. The modified Laspeyres index formula used in the
compilation of our Consumer Price Index is an example. Programming special
instances will, of course, continue to rely on the common computer lan-
guages---PL/1, FORTRAN, and COBOL. I expect we will also continue to rely on
the expert skills and energy of our systems analysts and programmers for much
of our data processing work long after we have generalized as much of our work
as possible.

The Plan in Practice

Well, what does all this mean? Given the picture of a central data base, a
data base management system and Data Documentation System, and blocks of
general-purpose, functional modules, what happens? It seems clear that for
future systems designers work will be reduced. The data base management software will help keep data in order. It will give assurance that corrections,
deletions, and additions are processed into and out of the data base expedi-
tiously and reliably. The general-purpose routines are available to provide
for data security and integrity. If data are new, they need be described
only once for the Codebook and Dictionary. All subsequent inquiries for data
descriptions may be satisfied by reference to the Data Documentation System
where the file descriptions are accurately portrayed and easily transcribed
(perhaps even automatically) into the relevant sections of new or modified
programs.

When an assignment calls for a statistical system from sample selection
through publication of results, system designers can consider the general-
purpose modules to address, edit, tabulate, and format the results for printing. They can provide for special analysis through the package of statistical
routines and display results on computer-drawn charts. The research economist, statistician, demographer, and other social scientist will have these
same tools available, and he need not be knowledgeable in the computer
science discipline. By naming the variables he wants tabulated and citing
the kinds of tabulations to be performed, the data base management system
will reach into the Codebook for the location of the values of the variable
cited, retrieve and deliver these figures to the tabulating system for processing, and display the results in the form expressed by the user.

The foregoing possibilities might seem unreal. Well, let's see. Most BLS
systems now use TPL for cross tabulation and some have used several of our

other building blocks to augment tailored programs. In one graphic instance,
an entire system was constructed from off-the-shelf items. Not a single line
of newly written code is used.   The system tests data for validity with a
routine from the collection of statistical analysis programs, updates and
extracts data with a prototype portion of the file manipulation system, man-
ages its data base with TOTAL, uses TPL and its Codebook for cross tabulation,
the Census X-11 program for seasonal adjustment, and PCL for table display.

In summary, the exploitation of this integrated and comprehensive system
depends on a willingness to seek new ways to solve old problems. The immense
potential of an open-ended computational system, attached to a large data
base of statistics representing a broad spectrum of the U.S. economy, is a
power that should be treated with respect. Economists, statisticians, other
social scientists, and systems analysts must operate as a team with communi-
cative skill to maintain the delicate balance between innovation and uncon-
trolled waste of funds, computer time, and analytical effort.

Discussion

Regionalization of statistical activities

Much of the discussion of the impact of the new technology on statistical
organizations related to the further regionalization of statistical activities. The desires of the regions for an increase in their own data activities has brought into focus the importance of the central/regional relationship. Two ways of establishing this relationship were discussed: 1) by
creating central data banks with regional access and 2) by computerizing the
regional planning at the regional level.

The advantage of greater participation of the regions in processing the data
is that it eliminates duplication and creates the possibility of a new
approach to statistics, building up local data to create national figures.
However, the disadvantage is that the passing of information back and forth
between the center and the regions by computers introduces the risk of unauthorized access to the data. Also, there is concern about the quality of
the data when it is created by autonomous local centers. It was suggested
that this latter aspect could be ameliorated by requiring regional data bases
to use standardized income and demographic data bases.

The technological activities of the countries with centrally-planned economies appeared to be similar to the experience of the market economy countries.
The U.S.S.R. delegate reported that they are building up autonomous data
banks in each local autonomous republic. There is much detailed information
at the local level; however, there is less data at the top. Another delegate
from a socialist country noted that by requiring all local units to use the
central register's standard methodology, they have kept the central office in
control.

Confidentiality

The discussion of this topic again emphasized the concerns over confidentiality. It will grow increasingly sensitive, especially if governmental involvement with regulation of individuals and enterprises increases, or will
recede if government reverses its role. However, most of the environmental
trends suggest that government intervention will increase and the problems
will become more sensitive even though the public will resist it. Concerns
over confidentiality are magnified because of the layman's impression that
anything available on computers in one place can be retrieved any other place.

Timeliness

In spite of its rapid advances, technology has not solved the problems of
timeliness that it might. As processing is only a small proportion of the
total time between collection and the final report, speeding up the processing time will have only a small impact on the total time.

One of the difficulties is that people are slow in providing the initial data
inputs. To speed up the initial receipt of the data, one country reported of
their experience in establishing a more personal relation and rapport with
the enterprises and then collecting the data by telephone instead of waiting
for it to come in the mail. This had improved the timely receipt of the
data, but it has increased the cost. Another participant suggested collecting the data by cables or tapes.

Suggestions for reducing the time between the collection of the data and the
beginning of processing included having the questionnaire coded at the same
time that it is completed by the interviewer so the questionnaire can be fed
into a machine already coded, checking the controls for the job before beginning processing to make sure that the job is correctly set up, working up
techniques to allow processing in batches and trying to distinguish between
production and test mode of programs. Big delays in programming occur when
people make changes in the program and the originator is not aware of this
until later. This could be eliminated by requiring changes to be made in
programs when they are in the test mode.

Processing and publication time could be saved by using data dictionaries
that have a time dimension which will avoid having to recode and build programs when definitions change, using generalized programs, and reducing the
preparation of tables by the use of photocomposition.

One country reported that they divided their data into two categories, those
in which the results were needed quickly and those that did not necessarily
require speed.

Concluding Remarks

The Environment for Statistics in the Coming Decade

The general environment in the coming decade will be very similar to the present environment. The problems central statistical offices will face will be similar to the problems they have at present. But the current trends will intensify and become more serious. In this environment, the statistician is going to have to fight to maintain his proper role in society and fight to insure that the public understands his role and function and cooperates for the good of society.

The coming decade will bring changes in the organization of statistics in most countries; these changes are clear. The changes will move toward more effective use of technology and greater developments in hardware and software. Statisticians will seek improved data dictionaries, standards, classifications systems, data linkage, and the use of registers to make the technology more useful and to have a more effective statistical program. There will be a trend to consolidate statistical activities with controls in better standards and procedures, and at the same time there will be a trend to decentralize some aspects of statistics so that statistics will be close to policy makers at the local and regional levels, as well as close to the policymakers at the Federal level.

There will be changes in the objectives of statistical offices, in their methods of operating and changes in the type of information base they develop. These changes are not so clear.

Before statisticians can define these changes in objectives, they are going to have to work hard to define priorities. There is no easy method of defining priorities. They will continue to reflect the judgments that statisticians have reached through negotiation with the governments and users they serve. The demand for improved statistical information will grow; the resources for statistics will decline and, even if they remain static, they will be inadequate to meet the demands. This will create more pressure for defining priorities. Chief statistical officers will not only have to develop a list of what is to be done, but a strategy of how it is to be done.

The Role of the Statistician and the Central Statistical Office in the Coming Decade

1. Statisticians must have a fundamental approach to the development and evolution of statistical organizations.

One of the participants of the Seminar had questioned whether chief statistical officers had their feet on the ground looking for facts or their heads in the clouds looking for new horizons with great imagination and innovation. This is the dilemma of central statistical offices: to be realistic and idealistic at the same time. Statisticians need the idealism to envisage an ideal statistical system and the realism to examine the current situation to see deviations from the ideal. They need a clear view of what is fundamental so they can examine current practices to see what needs to be repaired. The Canadian paper provides this fundamental approach to developing a statistical system.

This paper which lays the ground work for the future discussion of the evolution of statistical systems is fundamental and basic for several reasons. First, it is basic because it approaches statistical problems using systems analysis. Secondly, it reduces the statistical problems of concern to all countries to a common denominator for discussion. It is the heart of the whole statistical problem, the problem of statistical information freed from the social and economic context. It provides a basis from which one can build any statistical organization in any country. Thirdly, it is basic in that it provides a definition of the ideal statistical system from which deviations from the ideal can be determined; it gives a goal to be achieved in the future and a method of evaluating problems in the future; and it provides a checklist of how statisticians approach the next ten years.

Lastly, the paper is fundamental in looking at the future in that it goes back to the issues which were highlighted for the 80's - the statistical situation with respect to new resources, response burden, and the devolution of statistical activities--and asks "do we have the proper controls to control the whole system? In the non-market economy countries, there is more opportunity for line control. Countries of Western Europe are not going to have this line control over all elements so they will have to build up inducements to insure that their work is in the interest of all players, even those over which they do not have line control. The only way to do this is to develop some general fundamental principles and controls. The Canadian diagram may be essential for this purpose.

2. Central statistical offices will have to be more outward looking.

Central statistical offices will have to act in a wider context and cooperate with more actors in the information industry. Central statistical offices will have to have more contact with respondents to show them the role of statistics and help them associate the importance of having statistics and the benefits of following up on what is happening in society.

Central statistical offices must develop more contacts with users of statistics and develop a greater understanding of the analysis of data so that they can shape the statistics according to their uses.

There will be more producers of statistics outside of the central statistical system; central statistical offices must be ready to take the lead in coordinating the production of statistics.

Lastly, central offices need to improve their contacts with their governments in two directions. First, they must develop statistics that make sense to the policymakers. Second, they must develop contacts with the various records and registers growing in the various governments in order to coordinate them from the beginning and to supply all registers with a comparable infrastructure containing common nomenclatures, numbering systems, classification, etc. This will provide a means of extracting good and compatible statistical figures. The prestige and standing of central statistical offices will greatly depend on whether they take a leading role in governmental information systems as far as quantitative information is concerned.

3. Central statistical offices will have to determine what aspects of confidentiality are absolutely essential and crucial for statistical systems.

It will no longer be enough to just delete the names and addresses, particularly for small areas, because other organizations have information and can compare statistical information with their information and identify individuals or firms from "unidentifiable data." Statisticians will have to think through the issues and redefine confidentiality. They will be driven to this because administrative records will be used increasingly as a source of statistical data and because populations and firms are becoming increasingly sensitive about confidentiality.

4. Central statistical offices will have to fight for their role in maintaining statistical quality in administrative records.

They will have to show the public that statistical quality is important to maintain in any general information system. If statistical offices are not deeply involved in the way administrative files are processed, administrative authorities will take over and publish data that is meaningless because they do not care about the population coverage, the definition of the phenomena, etc. Confusion will arise in terms of the type of statistics that are published.

5. Chief Statisticians will have to be more concerned about human resources. Perhaps chief statisticians should study the diagram in the Canadian paper and ask what sort of people are in or operating these boxes. What sort of people should we have in the future? More mathematical statisticians? More substantive people? What sort of people should be in charge of EDP? What is the role of the specialist statistician?

6. Planning is absolutely crucial for central statistical offices.

Planning was particlarly emphasized in the discussion of priority setting. In the Eastern European countries priorities are set in terms of the economic plans. The actions in the market economy countries imply a plan, even though there may not be one. The job of chief statistical officers will be to try to articulate this implicit plan.

Many questions and gaps surround statisticians; many questions remain. The exchange of the Seminar will stimulate each of the countries to do much work on the issues and will comfort and help each country in learning that they all have common problems.

List of Participants

AUSTRIA

Mr. L. BOSSE — President, Central Statistical Office

Mr. J. LAMEL — Secretary, Economic and Social Advisory Council, Federal Economic Chamber

BELGIUM

Mr. P. F. VAN LANDEGHEM — Directeur General, Institut National de Statistique

BULGARIA

Mr. S. STANEV — President, Comite pour Information Sociale aupres du Conseil des Ministres

BYELORUSSIAN SSR

Mr. D. L. TCHERVANEV — Chief, Central Statistical Board of the Council of Ministers

CANADA

Mr. P. KIRKHAM — Chief Statistician, Statistics Canada

Mr. I. P. FELLEGI — Assistant Chief Statistician, Statistics Canada

Mr. G. LECLERC — Assistant Chief Statistician, Statistics Canada

Mr. J. RYTEN — Director General, General Statistics, Statistics Canada

Mr. W. DUFFETT — Vice President, Conference Board in Canada

Mr. B. PRIGLY — Coordinator (Statistician), International Relations, Statistics Canada

Mr. R. A. WALLACE — Assistant Chief Statistician, Census, Statistics Canada

CZECHOSLOVAKIA

Mr. J. KAZIMOUR — President, Federal Statistical Office

DENMARK

Mr. N. V. SKAK-NIELSEN — Head of Danmarks Statistik

FINLAND

   Mr. A. I. KENTTA   Director, Central Statistical Office

FRANCE

   Mr. E. MALINVAUD   Directeur General de l'Institut National de la Statistique et des Etudes
Economiques

GERMAN DEMOCRATIC REPUBLIC

   Mr. K. NEUMANN   Head of Research Center, Central Statistical Office

GERMANY, FEDERAL REPUBLIC OF

   Ms. H. BARTELS   President, Federal Statistical Office
   Mr. G. BURGIN   Chief of Subdivision, General Statistical Programmes and Organization
of Federal Statistics, Federal
Statistical Office

GREECE

   Mr. D. ATHANASSOPOULOS   Director General, National Statistical Service

HUNGARY

   Mr. J. BALINT   President, Central Statistical Office
   Mr. G. HORVATH   Head of Department, Central Statistical
Office
   Dr. V. FRANK   Adviser, Central Statistical Office

IRELAND

   Mr. T. P. LINEHAM   Director, Central Statistics Office

ITALY

   Mr. C. VITERBO   Manager of Studies Office, Instituto
Centrale di Statistica
   Mr. F. MAROZZA   EDP Manager, Instituto Centrale di
Statistica

LUXEMBOURG

   Mr. G. ALS   Directeur du Statec, Service Central
de la Statistique et des Etudes
Economiques

NETHERLANDS

   Mr. W. BEGEER   Adviser to the Director General,
                                         Central Bureau of Statistics

   Mr. C. A. OOMENS   Director, Economic Statistics,
                                           Central Bureau of Statistics

NORWAY

   Mr. O. AUKRUST   Director General, Central Bureau of
                                           Statistics.

POLAND

   Mr. T. WALCZAK   Vice President, Central Statistical
                                           Office

SPAIN

   Mr. R. BERMEJO   Deputy Director General of Coordina-
                                           tion, Instituto Nacional de
                                           Estadistica

   Ms. C. ARRIBAS   Head of International Coordination
                                           Instituto Nacional de Estadistica

SWEDEN

   Mr. I. OHLSSON   Director General, National Central
                                           Bureau of Statistics

SWITZERLAND

   Mr. C. A. MALAGUERRA   Chef de la Division de la Statistique
                                           Economique du Bureau Federal de
                                           Statistique

UNION OF SOVIET SOCIALIST REPUBLICS

   Mr. A. P. DRUCHIN   Chief, Central Statistical Board,
                                           Council of Ministers of the Russian
                                           Federation

   Mr. V. S. SHEVCHENKO   Economist, Foreign Relations Division,
                                           Central Statistical Board, Council
                                           of Ministers of the Russian Federa-
                                           tion

UNITED KINGDOM

   Sir CLAUS MOSER   Director, Central Statistical Office

UNITED STATES

   Mr. J. W. DUNCAN   Deputy Associate Director, Statistical
                                           Policy, Office of Management and
                                           Budget, Executive Office of the
                                           President

Mr. R. HAGAN   Acting Director, Bureau of the Census,
                                      Department of Commerce
Mr. J. SHISKIN   Commissioner, Bureau of Labor Statis-
                                      tics, Department of Labor
Mr. J. CARROLL   Assistant Commissioner, Office of
                                      Research and Statistics, Social
                                      Security Administration
Ms. M. ELDRIDGE   Administrator, National Center for Edu-
                                      cation Statistics, Department of
                                      Health, Education, and Welfare
Mr. W. E. KIBLER   Administrator, Statistical Reporting
                                      Service, Department of Agriculture
Ms. D. RICE   Director, National Center for Health
                                      Statistics, Department of Health,
                                      Education, and Welfare
Mr. A. H. YOUNG   Deputy Director, Bureau of Economic
                                      Analysis, Department of Commerce
Mr. R. PARKE   Social Science Research Council
Mr. W. SHAW   Consultant

YUGOSLAVIA

Mr. I. LATIFIC   Directeur, Office Federal de la Statis-
                                      tique
Mr. A. STANOJEVIC   Senior Adviser, Federal Statistical
                                      Office

PRESENT UNDER THE PROVISIONS OF PARAGRAPH 11 OF THE TERMS OF REFERENCE OF
THE ECONOMIC COMMISSION FOR EUROPE

ISRAEL

Mr. M. SICRON   Director, Central Bureau of Statistics

SPECIALIZED AGENCIES

UNESCO

Mr. G. CARCELES   Chief, Division of Statistics on Edu-
                                      cation, Office of Statistics

INTERNATIONAL MONETARY FUND

Mr. W. DANNEMANN   Assistant Director, Bureau of Statis-
                                      tics
Mr. R. KLINE   Chief, Data Fund Division, Bureau of
                                      Statistics

WORLD BANK

Mr. M. E. MULLER   Director of Computing Activities Dept.

INTERGOVERNMENTAL ORGANIZATIONS

EUROPEAN ECONOMIC COMMUNITY

Mr. G. W. CLARKE   Adviser to the Director General of the
                                          Statistical Office

NON-GOVERNMENTAL ORGANIZATIONS

INTER AMERICAN STATISTICAL INSTITUTE

Mr. T. H. MONTENEGRO   Secretary General

SECRETARIAT

ECE STATISTICAL DIVISION

Mr. B. DAVIES
Mr. R. GENTILE
Mr. A. KAHNERT

UN STATISTICAL OFFICE

Mr. S. A. GOLDBERG